"Paula Hanasz not only provides a very careful analysis of factors responsible for lack of effective cooperation in the Ganges-Brahmaputra-Meghna basin among the riparian countries, but also suggests some thoughtful ways forward. Anyone interested in transboundary water governance in South Asia will treasure this book."

– *Ashok Swain, Dept of Peace and Conflict Research, Uppsala University, Sweden*

Transboundary Water Governance and International Actors in South Asia

International organisations such as the World Bank began to intervene in the transboundary water governance of the Ganges–Brahmaputra–Meghna river basin in the mid-2000s, and the South Asia Water Initiative (SAWI) is its most ambitious project in this regard. Yet neither SAWI nor other international initiatives, such as those of the Australian and UK governments, have been able to significantly improve transboundary water interaction between India, Nepal, Bhutan and Bangladesh.

This book identifies factors that contribute to water conflicts and that detract from water cooperation in this region. It sheds light on how international organisations affect these transboundary water interactions. The book discusses how donor-led initiatives can better engage with transboundary hydropolitics to increase cooperation and decrease conflict over shared freshwater resources. It is shown that there are several challenges: addressing transboundary water issues is not a top priority for the riparian states; there is concern about India's hydro-hegemony and China's influence; and international actors in general do not have substantial support of the local elites. However, the book suggests some ways forward for improving transboundary water interaction. These include: addressing the political context and historical grievances; building trust and reducing power asymmetry between riparian states; creating political will for cooperation; de-securitising water; taking a problemshed view; strengthening water sharing institutions; and moving beyond narratives of water scarcity and supply-side solutions.

Paula Hanasz has completed a PhD at the Australian National University, Australia. She consults on issues of defense, national security and non-traditional security threats.

Earthscan Studies in Water Resource Management

Water, Knowledge and the Environment in Asia
Epistemologies, practices and locales
Edited by Ravi Baghel, Lea Stepan and Joseph K.W. Hill

Community Management of Rural Water Supply
Case studies of success from India
Paul Hutchings, Richard Franceys, Stef Smits and Snehalatha Mekala

Drip Irrigation for Agriculture
Untold stories of efficiency, innovation and development
Edited by Jean-Philippe Venot, Marcel Kuper and Margreet Zwarteveen

Water Policy, Imagination and Innovation
Interdisciplinary approaches
Edited by Robyn Bartel, Louise Noble, Jacqueline Williams and Stephen Harris

Rivers and Society
Landscapes, governance and livelihoods
Edited by Malcolm Cooper, Abhik Chakraborty and Shamik Chakraborty

Transboundary Water Governance and International Actors in South Asia
The Ganges–Brahmaputra–Meghna basin
Paula Hanasz

The Grand Ethiopian Renaissance Dam and the Nile Basin
Implications for transboundary water cooperation
Edited by Zeray Yihdego, Alistair Rieu-Clarke and Ana Cascao

Freshwater Ecosystems in Protected Areas
Conservation and management
Edited by Max C. Finlayson, Jamie Pittock and Angela Arthington

For more information and to view forthcoming titles in this series, please visit the Routledge website: www.routledge.com/books/series/ECWRM/

Transbaoundary Water Governance and International Actors in South Asia

The Ganges–Brahmaputra–Meghna Basin

Paula Hanasz

Routledge
Taylor & Francis Group
LONDON AND NEW YORK

earthscan
from Routledge

First published 2018
by Routledge

2 Park Square, Milton Park, Abingdon, Oxfordshire OX14 4RN
52 Vanderbilt Avenue, New York, NY 10017

Routledge is an imprint of the Taylor & Francis Group, an informa business

First issued in paperback 2019

British Library Cataloguing-in-Publication Data
A catalogue record for this book is available from the British Library

Library of Congress Cataloging-in-Publication Data
A catalog record for this book has been requested

ISBN: 978-1-138-09754-4 (hbk)
ISBN: 978-0-367-35223-3 (pbk)

Typeset in Goudy
by Out of House Publishing

Contents

Acknowledgements ix
Acronyms and abbreviations x

Introduction 1

1 International actors in the resolution of transboundary water conflicts 18

2 Legacies and challenges for water governance in the Ganges–Brahmaputra problemshed 32

3 The transboundary water governance agenda in the Ganges–Brahmaputra problemshed 45

4 Why are international actors interested in the Ganges–Brahmaputra problemshed? 59

5 Normative claims of international actors 67

6 From water conflict to water cooperation through the World Bank? 81

7 Mixed reactions to foreign-led initiatives from riparian stakeholders 91

8 Obstacles to positive water interaction in the Ganges–Brahmaputra problemshed 101

9 The small issue of the big brother: the effect of India's hydro-hegemony 115

10 Limitations of foreign-led approaches 122

11 Structural challenges in the global system of foreign aid 136

12 Insights for international actors in the Ganges–Brahmaputra problemshed and beyond 143

Conclusions 156

Glossary 169
References 173
Index 189

Acknowledgements

First and foremost, I would like to thank everyone who made the time and effort to speak with me about the hydropolitics of the Ganges–Brahmaputa–Meghna basin over the years, and especially during my PhD fieldwork in 2014. Some people preferred to remain anonymous, but my heartfelt thanks goes out to them, as well as to: Ajaya Dixit, Amita Bhaduri, Anjal Prakash, Anumita Raj, Arunabha Ghosh, Ashok Swain, Bill Young, Brian Cook, Bushra Nishat, C. Raja Mohan, David Grey, David Molden, Dhaval Desai, Dipak Gyawali, Doug Hill, Dunu Roy, Gareth Price, Gary Jones, Golam Rabbani, Hamsa Iyer, Himanshu Thakkar, John Dore, Joydeep Gupta, Mandakini Devasher Surie, Mihir Bhonsale, Pranab Kumar Ray, Ramaswamy R. Iyer, Ramesh Bhushal, Rishi Aggarwal, Rohan D'Souza, Russell Rollason, Ryan Thew, S. Vishwanath, Sabita Kaushal, Sagar Prasai, Samir Saran, Shafiqul Islam, Shawahiq Siddiqui, Shawn Novel, Sonali Mittra, Srinivas Chokkakula, Sujeev Shakya, Tahira Syed and Tira Foran.

For their friendship, advice and support in the world of transboundary water studies, I would like to thank: Florian Krampe, Francesco Sindico, James Bond, Kevin Wheeler, Marian Neal (Patrick), Mark Zeitoun, Naho Mirumachi, Peter Mollinga, Rebecca Peters, Remy Kinna and Rozemarijn ter Horst. Thanks also to Daniel Connell.

This book would not have been possible without the tireless support and encouragement of Tim Hardwick, Amy Johnston and the rest of the wonderful team at Routledge/Earthscan. Thank you all!

Acronyms and abbreviations

ADB	Asia Development Bank
ADD	Abu Dhabi Dialogue
ADDG	Abu Dhabi Dialogue Group
AIIB	Asia Infrastructure Investment Bank
AusAID	Australian Agency for International Development (part of DFAT since 2013)
AWP	Australian Water Partnership
bcm	billion cubic metres
CEEW	Council on Energy, Environment and Water
CSIRO	Commonwealth Scientific and Industrial Research Organisation (Australia)
DFAT	Department of Foreign Affairs and Trade (Australia)
DFID	Department for International Development (United Kingdom)
GBM	Ganges–Brahmaputra–Meghna mega-basin
GIZ	Gesellschaft für Internationale Zusammenarbeit (Germany)
GLOF	glacial lake outburst flood
GSBA	Ganges Strategic Basin Assessment
GWP	Global Water Partnership
ICIMOD	International Centre for Integrated Mountain Development
INGO	International Non-Government Organisation
IUCN	International Union for the Conservation of Nature
IWMI	International Water Management Institute
IWRM	Integrated Water Resources Management
MDBA	Murray-Darling Basin Authority
MDG	Millennium Development Goals
MW	mega watt
NBI	Nile Basin Initiative
NGO	Non-Governmental Organisation
NWP	National Water Policy
RBO	River Basin Organisation
SAARC	South Asian Association for Regional Cooperation

SaciWATERs	South Asia Consortium for Interdisciplinary Water Research Studies
SAWGP	South Asia Water Governance Programme
SAWI	South Asia Water Initiative
SDG	Sustainable Development Goals
SIWI	Stockholm International Water Institute
SMEC	Snowy Mountain Engineering Corporation
TWINS	Transboundary Waters Interaction Nexus
UNWC	United Nations Convention on the Law of the Non-Navigational Uses of International Watercourses (1997)
WBG	World Bank Group
WCD	World Commission on Dams
WWF	World Wildlife Fund

Introduction

The fear of water wars, and the understanding that water can be a stress multiplier in socio-political conflict at all levels, has generated in recent years some impetus for increasing water cooperation between riparians (the countries that are connected by a common river system). How to improve transboundary water cooperation (that is to say, the mutually agreeable governance of shared freshwater resources between states) has been of particular interest to policymakers, aid donors and scholars of conflict. The prevailing idea is that cooperation arises from good governance. As Araral and Wang (2013, p. 3945) have shown, '[w]ater scholars, policy makers and donors over the last decade generally agree that improving water governance holds the key to solving water insecurity in developing countries'. But who is to be in charge of improving water governance? Is it the riparian states themselves? Or neutral third parties with international experience of facilitating hydro-diplomacy?

The case of the Ganges–Brahmaputra problemshed provides some answers to these questions. 'Fine minds in South Asia are seeking solutions for the looming water and river crisis in this region' argues Colopy (2012, p. 319), and '[s]ome outsiders would like to help ... Perhaps South Asia could use help from outsiders, perhaps not; but governments in the rest of the world are starting to pay attention.' But the effect of foreign-led interventions on the water interaction between India–Nepal, India–Bhutan, and India–Bangladesh illustrates that transboundary water cooperation cannot simply be imposed from the outside. Instead, the experience of the World Bank and other aid donors in the Ganges–Brahmaputra problemshed suggests that transboundary cooperation is a complex phenomenon, which can only emerge when multiple conditions are met – and even then it cannot be guaranteed.

At first glance this may not be an especially encouraging finding for the growing number of scholars and practitioners who believe that increasing cooperation between water stakeholders at all scales is the key to reducing a slew of water-related risks, including transboundary water conflicts. After all, if transboundary water cooperation cannot be guaranteed, even with the best intentions, best practice and best theoretical underpinnings, is there any role for foreign-led interventions at all? Are transboundary water basins with simmering conflicts and weak or ineffectual governance fated to muddle along indefinitely?

I do not question here whether or not international organisations and foreign aid donors should be intervening in transboundary water governance, either in the Ganges–Brahmaputra problemshed or elsewhere – there are many reasons why they do and will likely continue to do so. Nor do I question whether transboundary water cooperation is a worthwhile goal for international actors to pursue generally (or, indeed, whether it is a desirable outcome for all riparians). Rather, I examine the normative claims that international organisations make in regard to water governance, and critique the approaches used to pursue them. I compare the effect international organisations and foreign aid donors have had in the Ganges–Brahmaputra problemshed with their stated objectives of increasing transboundary water cooperation and decreasing conflict. This analysis is based on an in-depth investigation of perspectives, experiences and needs from key stakeholders within the problemshed.

Taking a transdisciplinary approach such as this is unique and significant because most evaluations of transboundary water governance focus on international expertise rather than that of those within the basin or problemshed in question. Julien (2012, p. 62) argues that 'hydropolitics is what societies make of it'. By examining what affected societies make of transboundary water interaction in the Ganges–Brahmaputra problemshed, this book sheds new light on a complex problem that has hitherto been largely examined through the lens of international expertise. Yet, as Julien (2012, p. 62) points out, 'it is the political dynamics inside which transboundary water management occurs that must be the focus of analysis'.

With this approach in mind, I examine from the local perspective the levels of transboundary water conflict and cooperation – together known as water interaction – between India–Nepal, India–Bhutan and India–Bangladesh. Together, these countries form the Ganges–Brahmaputra problemshed. This problemshed lies in the Ganges–Brahmaputra–Meghna basin, in which China is also a riparian. Yet the centre of gravity for foreign-led interventions is contained within the political, not hydrological, boundaries and thus the concept of problemshed is more appropriate for limiting the scope of this book than is the concept of the watershed. The concept of the problemshed is used herein because, as Mollinga (2008b, pp. 7–8) points out, water resources management stretches well beyond the physical boundaries of these units, and span the globe and history.

I argue that because of the complex nature of transboundary water governance, and the inherent unpredictability of complex adaptive systems, external actors alone are unable to directly bring about positive water interaction between riparians. They may, however, to some extent influence the conditions from which cooperation may emerge. What those conditions are, and how they could be influenced, are the subject of my research.

The rest of this Introduction tells the story of the foreign-led interventions in the Ganges–Brahmaputra problemshed, and provides an overview of how these interventions have been experienced within the region. These insights deepen the understanding of transboundary conflict and cooperation, and also provide some approaches for improving transboundary water interactions.

Why are transboundary water interactions in the Ganges–Brahmaputra problemshed a problem?

The Hindu Kush Himalayan region, of which the Ganges–Brahmaputra problemshed is a part, has one of the largest bodies of ice outside the polar caps, and the mountains are often referred to as the 'water towers' of Asia. These mountains are the major source of both surface and groundwater in the region during the dry season, and have the potential to play a vital role in energy security; if properly harnessed, hydropower could transform the lives of the people living in these river basins. But, as Vaidya and Sharma (2014, p. 3) explain, the contributions of meltwater and rainfall coincide during the monsoon, creating a situation of too much water in the wet season and too little water in the dry season. This situation is exacerbated by the lack of water storage facilities. They show that the increasing gap between water availability and water demand outside the monsoon season is already posing a serious threat to livelihoods and economic development in South Asia.

The Ganges–Brahmaputra problemshed is an area of plenty *and* of scarcity. Both extremes are likely to be exacerbated by the effects of climate change. As Hill (2015, p. 739) notes, '[t]he phrase "too much and too little water" has been used as a shorthand for the challenges that climate presents for the Hindukush Himalayas, with droughts and floods becoming more commonplace'.

The Ganges–Brahmaputra–Meghna basin is one of the richest basins in the world in terms of natural resources (hydropower, fisheries, forestry, irrigated agriculture, navigation, environmental amenities, tourism, minerals, etc.) Yet India, Nepal and Bangladesh are some of the poorest in the world and sometimes called 'the poverty triangle' (Dinar et al. 2007, p. 251). It is also a densely populated region. The greater Hindu Kush Himalaya is home to nearly half of the world's population; approximately 630 million people in an area of 1.7 million square kilometres (Vaidya & Sharma 2014, p. 17).

Water storage is a particularly important issue for transboundary governance because it can be used to even out water supply between where it is comparatively plentiful – Nepal and Bhutan – and where it is scarce – India. In addition to flow augmentation for downstream irrigation, Nepal's storage potential of 88 bcm could bring other mutual benefits, such as hydropower generation, flood control during the monsoon and flow augmentation for navigation (Onta 2001, p. 106).

Although the issue of hydropower generation in the Ganges–Brahmaputra problemshed is the one most often – and most vigorously – debated within the region and internationally, it is water for irrigation that has broader effects in terms of human security, livelihoods and food production. As Dinar et al. (2007, p. 254) point out, '[a]griculture accounts for nearly one half of all freshwater usage in the [Ganges–Brahmaputra–Meghna] basin, making water supply one of the most significant barriers to economic development'. Agriculture, in turn, is vulnerable to drought and flooding.

These issues of hydropower generation, irrigation, flooding and sedimentation are brought together in the prevailing narrative of freshwater resource development in the region. The current thinking is that there is almost infinite hydropower potential in the Himalayas (especially Nepal) and that it should be exploited for the benefit of all riparians. Planned upstream dams are framed as 'multi-purpose' and promise to bring multiple benefits: economic benefit for Nepal, energy security for northern India, water storage for dry season flow augmentation and irrigation, and flood control throughout the basin. This narrative is based on the dominant paradigm of water management: command and control of nature, and the pursuit of technological and engineering solutions to water related problems.

In addition to these issues, water interactions between India, Nepal, Bhutan and Bangladesh are not positive (in the sense of being mutually beneficial and satisfactory). As Dinar et al. (2007, p. 248) note in regard to the Ganges basin, it 'has tremendous joint development potential that has not yet been realised by its riparian states. Rather, the Ganges Basin is more popularly known for its rich history of disputes.' Factors in this 'rich history of disputes' include the lack of strong regional identity, the securitisation of water issues, the presence of a hydro-hegemon that struggles to be a regional leader, the reliance on bilateral rather than multilateral approaches to transboundary water governance, and a number of unresolved disputes between India–Nepal and India–Bangladesh. Wirsing et al. (2013, p. 14) also add faltering river diplomacy, and simmering separatist tensions and political turmoil to these conflict-inducing factors.

Uprety (2014) summarises the transboundary water interactions in South Asia by saying that while some riparian states claim to be in favour of cooperative, multinational approaches to water governance, collective multilateral attention to the problems of transboundary waters has been rare. Historically, he argues, the region has lacked a collective strategy and bilateralism remains the focus (Uprety 2014). Riparian states engage only half-heartedly in transboundary water initiatives, while in practice only advancing those strategies that serve their own national interest. This is, argues Uprety (2014), due to their mutual distrust, their inability to delink hydropower from water resources, and their reluctance to establish a precedent for multilateralism.

As Asthana and Shukla (2014, pp. 275–277) show, there is also no sense of collective action for dealing with non-traditional security threats, such as water conflicts. Indeed, they argue that '[n]on-traditional threats are also securitised across borders due to a bitter past, mistrust and hatred that dominates the socio-political structure in these countries'. The securitisation of water continues to be a driver of negative water interaction in the Ganges–Brahmaputra problemshed.

India is by far the most powerful state in this area and as such is able to dominate the transboundary water governance agenda. This creates tensions and mistrust between riparian states. As Hill (2008, p. 75) shows, '[t]he geopolitics of the region is arguably defined by the dominance of India and issues surrounding whether trans-boundary disputes should be handled bilaterally or internationalised continue to provoke tensions' and furthermore these tensions 'are likely to intensify as demand for water becomes more acute in the future'.

It is the possibility of these tension intensifying that is one of the primary drivers of foreign-led interventions in the Ganges–Brahmaputra problemshed. The Ganges–Brahmaputra problemshed is particularly vulnerable to water conflicts. Water conflicts are generally understood to be the result of state fragility (recent violence, obstacles to economic development, weak management institutions), and high water stress (National Research Council of the National Academies 2012, p. 92). As water governance issues gain higher political attention locally, nationally, regionally and globally, cooperation and hydro-diplomacy are increasingly seen as means of attenuating water-related disputes, particularly in transboundary settings (Keskinen et al. 2014, p. 39).

As Earle et al. (2015, p. 41) note, there is currently no paucity of collaborative efforts in South Asia that explore effective response strategies to climate change and other challenges of our era. Some of these collaborative efforts are an extension of existing initiatives or engagements by international organisations and foreign aid donors in the region. For example, the governments of Australia and the United Kingdom have begun focusing their diplomatic and foreign aid relations in South Asia on climate change, water security, sustainable development and other concerns of the age. Other collaborative efforts are entirely new and specific to transboundary water governance. The most significant of these are outlined below.

The Abu Dhabi Dialogue and the South Asia Water Initiative

The confluence and culmination of global challenges such as climate change, water scarcity and other non-traditional security threats, and regional challenges for water governance (described in Chapters 2 and 3), led in the mid-2000s to an international round of talks called the Abu Dhabi Dialogue (ADD).

The ADD was funded and convened by the World Bank (Earle et al. 2015, p. 41), and periodically brought together senior members of government, academia and civil society from Afghanistan, Bangladesh, Bhutan, China, India, Nepal and Pakistan in a non-formal, consultative process intended to foster regional cooperation on various challenges facing Himalayan rivers (SAWI 2015c). Earle et al. (2015, p. 41) describe the ADD as 'the only multilateral framework involving all the riparian countries of the Ganges–Brahmaputra basin that is engaged in an informal and consultative process' and they point out that the ADD aimed to build trust between riparians primarily through information exchange and the co-management of development projects.

The ADD was instituted with a decade-long agenda beginning in 2006 (Vaidya & Sharma 2014, p. 4). In 2007, the seven countries met to 'share global experience on international waters and benefit-sharing, with the goal of achieving constructive convergence on common challenges' (SAWI 2015d). For the World Bank, however, the ADD was a chance to create a 'common water-sharing framework in line with the Nile Basin Initiative' (Earle et al. 2015, p. 41). Yet, the Nile Basin Initiative (NBI) has not been an unmitigated success in terms of increasing transboundary water cooperation (Swain 2011, p. 696). The ADD can therefore be seen as an opportunity to transpose the successes, such as they were, of the Nile Basin Initiative into a different region, and to improve on the failures. The Ganges–Brahmaputra was an appropriate choice as there was – and still is – no other basin-wide collaborative process, yet the potential for water conflicts remains high.

The 2007 meeting of the ADD was also when the South Asia Water Initiative (SAWI) was established. The mandate of SAWI is to 'promote the goals of poverty reduction, economic growth, mitigation and adaptation to climate change, water security and regional peace through significant and measurable improvements in water resources management and development' (SAWI 2015d). The main objective of SAWI is to promote regional cooperation and the sustainable management of water resources of the Himalayan rivers based on the principles of Integrated Water Resource Management (IWRM), inclusive and multi-disciplinary dialogue, and participatory and deliberative processes (SAWI 2015a).

Another key principle of SAWI is to deliver mutual benefits (SAWI 2015a). The theory of change is that mutual benefits accrue

> from coordinated planning and joint investment for new infrastructure that ensure equitable upstream–downstream sharing of both benefits and impacts, and from coordinated planning and joint investment into upstream watershed rehabilitation that deliver both local benefits and downstream water quality benefits.
>
> (SAWI 2015b)

In other words, benefit sharing is linked in SAWI methodology to transboundary water cooperation.

It is worth noting that these principles and objectives – IWRM, deliberative governance, benefit sharing – are all driven by global narratives around international development and environmental management (these narratives are discussed in Chapter 5). They have not come from within the Ganges–Brahmaputra problemshed, or been put on the transboundary water governance agenda by riparian stakeholders.

A further noteworthy aspect of SAWI is that it is funded exclusively by external donors: Australia, the United Kingdom and Norway (World Bank 2014b, pp. 30–34). The countries in which SAWI operates (that is, the Abu Dhabi Dialogue Group member states) do not make even symbolic contributions. This illustrates a significant distinction between the ADD and SAWI. While the ADD is about the riparian states of the Indus, Ganges and Brahmaputra basins driving dialogue (facilitated by the World Bank) about shared water issues, under SAWI it is the donors – particularly the World Bank, Australia and the United Kingdom – setting the agenda. The riparian states remain more or less passive recipients of the initiative.

Insights from within the problemshed

The year 2016 marked the end of the decade-long ADD process. What can we learn from it? How do the experiences of stakeholders within South Asia reflect on World Bank initiatives into transboundary water governance? What are the lessons for foreign-led interventions generally, and what options remain for improving transboundary water interaction in the Ganges–Brahmaputra problemshed specifically?

Earle et al. (2015, p. 41) remark that the World Bank failed in its hope to bring South Asian countries within a common water-sharing framework. But just as the ADD was an opportunity to improve on the Nile Basin Initiative, so too there are lessons to be learned from the ADD. These lessons can be applied to whatever comes next in the Ganges–Brahmaputra problemshed, but also to other basins or regions of the globe where the World Bank – or other international organisations or foreign aid donors – may be planning similar ventures. There are four main observations about how international organisations and aid donors conduct interventions into transboundary water governance, and how this affects their effectiveness in terms of increasing cooperation between riparian states.

First, the ADD and SAWI have met with only middling support from riparians, especially India, and this, in Asthana and Shukla's words (2014, p. 344), makes 'multilateralism or third-party interventions in a regional waterscape … not only difficult but also impossible'. Second, there are limitations inherent in the way that international actors approach transboundary water governance. As one analyst put it, there are 'a lot of solutions looking for problems' in the Ganges–Brahmaputra problemshed, and that 'when all you have is a hammer, then every problem becomes a nail' (Anonymous 10, 2014). Third, the system in which international organisations and foreign aid donors operate – the system of foreign aid – is crowded, inefficient, short-sighted and driven by complex agendas.

Furthermore, the effectiveness and impact of these actors in the transboundary water governance of the Ganges–Brahmaputra problemshed will likely never be known because of the twin problems of effect and attribution (Mayne 1999, pp. 2–3). The crowded and complex system of foreign aid means that it is almost impossible to separate out the effects of individual donors and external factors on common objectives (Riddell 2014, p. 26). And the desired effects can be brought about by factors entirely unrelated to the project or programme working towards them (Riddell 2014, p. 45).

Nonetheless, there are ways in which international actors may be able to mitigate, if not entirely overcome, these challenges. Riparian stakeholders, however, have unique insights into how transboundary water conflict may be reduced, and cooperation increased, in the Ganges–Brahmaputra problemshed. These include: addressing the political context and historical grievances; building trust and addressing power asymmetry between riparian states; creating political will for cooperation; de-securitising water; focusing on water justice outcomes; taking a problemshed view; strengthening water sharing institutions; and moving beyond narratives of water scarcity and supply-side solutions. Incorporating these approaches into the foreign-led intervention in the Ganges–Brahmaputra problemshed, as well as future interventions into other problemsheds, may help create the conditions in which positive transboundary water interactions may emerge.

This book argues that the complexity of transboundary water interaction prevents any one actor from definitively increasing cooperation between riparian states. There may, nonetheless be a circumscribed role for international organisations and foreign aid donors in supporting riparian states to foster the conditions in which cooperation may emerge. These are discussed throughout.

Overview of chapters

Chapter 1 examines the role of international actors in transboundary water conflict, cooperation and interaction. First, it discusses the evolution of engaging third parties in the resolution of international disputes. This includes an overview of the traditional roles of the third party – as mediator, conciliator, arbitrator and as adjudicator. A critique is provided of this functionalist understanding of international conflict: the context of culture and power dynamics between parties is often overlooked, and may be an obstacle to cooperation that third parties simply cannot redress.

Second, this chapter discusses how international actors have come to be involved in transboundary water disputes. Often this role involves facilitating Track II dialogue, or other forms of hydro-diplomacy. International organisations and aid donors have also been involved in studies or basin-wide assessments that outline possibilities for benefit sharing. This chapter notes, however, that benefit sharing outcomes are more often a *consequence* – rather than a cause – of cooperative water governance. The chapter concludes with a discussion about the circumscribed role of the third party in resolving international water conflicts.

The book then moves on to focus on the geographic context of South Asia. Chapter 2 describes the legacies and challenges for water governance in this region. These include colonial approaches to the control of rivers, and the current narratives of water 'scarcity' that are often addressed through supply-side solutions. This chapter also discusses how global challenges – rapid population growth and urbanisation, climate change, food security – play out in regard to transboundary waters.

Chapter 3 outlines the main issues on the water governance agenda in the Ganges–Brahmaputra problemshed. These include upstream water storage, hydropower generation, irrigation, flooding and sedimentation. These are the most important and controversial issues between the riparian states, as well as the issues receiving the most attention from international actors involved in water governance here. Each is discussed in turn, first from the perspective of stakeholders within the problemshed, then from the perspective of the World Bank.

The next three chapters shift to a focus on the international organisations and aid donors engaging in transboundary water governance in the Ganges–Brahmaputra problemshed. Chapter 4 asks: why are international actors interested in this area? It explains the increasing interest in the socio-economic development and political stability of South Asia, as well as the waning influence of the World Bank in the region. The construction of water infrastructure is no longer the approach for water resources development favoured by the World Bank or, indeed, stakeholders in developing countries. Instead, a more nuanced approach is increasingly preferred; an approach of good governance and benefit sharing outcomes.

Chapter 5 examines the normative claims made by international actors in regard to water governance. It discusses the ways in which global water governance paradigms and practices have changed in recent decades. There has been a growing emphasis on 'good governance' and water justice. At the same time, a tension exists between the long-established 'command and control' approach to environmental resource management versus ecological principles. This chapter also discusses the emergence of non-traditional security threats, especially the risks associated with a changing climate. Water security is an increasingly important concern for international development, as are sustainable development and poverty eradication.

Based on these normative claims, Chapter 6 looks at how the World Bank and associated donor processes are addressing transboundary water disputes and attempting to improve water interactions between India, Nepal, Bhutan and Bangladesh. The main goals that underpin the work of these international actors include good water governance and hydro-diplomacy, regional stability, sustainable development, poverty eradication and climate change resilience. Knowledge sharing is a fundamental task underpinning all these goals.

The second half of the book turns to the challenges and opportunities for international actors working to improve transboundary water governance in the Ganges–Brahmaputra problemshed and beyond. Chapter 7 examines the attitudes of the riparian stakeholders towards these foreign-led initiatives. On the whole, reactions are mixed. There is some resentment about top-down approaches, but also a reluctant acceptance that neutral third parties with relevant expertise may be needed to improve transboundary water governance. These mixed reactions limit the effectiveness of international actors because there remains low political will for addressing shared water issues.

Further obstacles to transboundary water cooperation are discussed in Chapter 8. There is an underdeveloped civil society and weak regionalism. At the same time, South Asian states see collective action as a threat to sovereignty. As such, zero-sum attitudes to water sharing prevail, and transboundary water issues are securitised. A culture of secrecy prevails around hydrological data, and this impedes trust and cooperation between riparian states. All these factors make the work of third parties more challenging – and all the more important.

At the same time, India's hydro-hegemony creates further mistrust and resentment among the other riparian states. This is the subject of Chapter 9. It first discusses the perceptions of Nepal, Bhutan and Bangladesh towards India, then examines them against the theory of hydro-hegemony. India is neither a leader, nor a bully; but India's insistence on strictly bilateral approaches to transboundary water governance does impede cooperation. The chapter concludes with a discussion about how to counter the hydro-hegemon.

Chapter 10 examines limitations of initiatives led by international actors. Knowledge sharing, though important, is not the silver bullet for all transboundary water issues in the Ganges–Brahmaputra problemshed. Likewise, trying to address water problems on a basin-wide scale may not always lead to the most appropriate solutions. Furthermore, international actors tend not to account for the political realities of the region, but assume that increasing transboundary water cooperation will result in greater overall socio-political stability: this is not necessarily the case. There are, lastly, limits to Track II dialogue, especially over the short term.

International interventions tend to have relatively short funding cycles. This is a problem for realising transformational change, such as improving transboundary water interactions. This, and other structural challenges of foreign aid are discussed in Chapter 11. The global system of international development is crowded, inefficient and risk-averse. Then there are the problems of effect and attribution in evaluating effectiveness, especially over the long term. Foreign aid is also driven by foreign agendas rather than local considerations. These are all challenges for international actors working on transboundary water governance in the Ganges–Brahmaputra problemshed.

All these limitations discussed in previous chapters are a challenge for international actors, but not insurmountable. Chapter 12 examines some ways in which foreign-led interventions into transboundary water governance in this region and beyond could be improved. Addressing historical grievances and the political context is certainly important. Building trust and political will, as well as addressing power asymmetry between riparian states is also critical. Focusing on water justice outcomes rather than benefits sharing outcomes may also prove productive, as could moving beyond narratives of scarcity and supply-side solutions. Thinking in terms of a problemshed, rather than the whole basin, may facilitate the development of more contextually appropriate policies. International actors could also continue strengthening water sharing institutions and lead by example in moving beyond narratives of water scarcity and supply-side solutions.

The final chapter summarises the current state of transboundary water interactions in the Ganges–Brahmaputra problemshed. It revisits the reasons for and approaches of foreign-led initiatives here. This chapter draws out conclusions about the role of international actors in addressing water governance issues in developing countries.

Definition of key terms

International actors, foreign aid donors and third parties

'International actors' is a term used herein as a catch-all for international and transnational organisations, foreign aid agencies, states acting on the international stage or pursuing their interests abroad, and international non-government organisations (INGOs). In the context of the Ganges–Brahmaputra problemshed, these actors are almost exclusively western and in the main aligned with a liberal-democratic ideology. The terms 'external actors' and 'third party' are used interchangeably throughout to emphasise the outsider role that these actors play vis-à-vis riparian stakeholders. However, while all international organisations are third parties, not all third parties are international actors.

A third party is '[a] person or group besides the two primarily involved in a situation, especially a dispute' according to the Oxford Dictionary. In the Ganges–Brahmaputra problemshed, it is any individual, government, organisation or other group that does not come from India, Nepal, Bhutan or Bangladesh but that nonetheless is engaged in water governance here.

The most significant international actors of relevance here are the World Bank, the South Asia Water Initiative (SAWI, led by the World Bank), the United Kingdom (which contributes to the SAWI but also funds the South Asia Water Governance Programme through the Department for International Development), and Australia (another major contributor to SAWI).

Norway and Japan are also aid donors in this space, but their contribution is significantly smaller and therefore less relevant here. Likewise, there are a number of international and multinational initiatives that are each only tangentially related to transboundary water cooperation but all together contribute to the bigger picture of foreign-led interventions. The International Centre for Integrated Mountain Development (ICIMOD), for example, is an important multinational organisation working primarily on transboundary collaborations on scientific knowledge production. Hill (2015, p. 740) describes ICIMOD as 'a regional intergovernmental learning and knowledge-sharing centre based in Kathmandu that serves the member countries of the Hindukush Himalayas and has an incredibly significant role in capacity building in the region', and argues that it is '[p]erhaps the best-known pan-South Asian agency in terms of water and climate change issues'.

Other international organisations that also affect transboundary water interaction in the Ganges–Brahmaputra problemshed include, but are not limited to, the Global Water Partnership (GWP), the International Water Management Institute (IWMI), the International Union for the Conservation of Nature (IUCN), World Wildlife Fund (WWF), the Asia Development Bank (ADB) and the Asia Foundation. The German foreign aid agency, Gesellschaft für Internationale Zusammenarbeit (GIZ), is also active in the Ganges–Brahmaputra problemshed. These, and other internationally funded or run organisations, are referred to herein as foreign-led interventions unless stated otherwise.

An international actor can become engaged in a situation in different ways and for different reasons. In terms of transboundary water governance, international organisations and foreign aid donors intervene with the permission of riparian states. A riparian state may or may not initiate a request for assistance from a third party, but these external actors are generally accepted and largely seen as legitimate – even if at the same time their presence is resented (attitudes toward foreign-led interventions in the Ganges–Brahmaputra problemshed are discussed in Chapter 7). The main normative reason for international actors intervening in transboundary water interaction is to improve governance and increase cooperation, although there are other implicit or unstated drivers as well (these are examined in Chapters 4 and 5).

Problemshed

A problemshed is 'a geographic area that is large enough to encompass management issues, but small enough to make implementation feasible' (Islam & Susskind 2013, p. 321). The problemshed, then, has both a physical – geographic – dimension, as well as a socially constructed one. As Mollinga et al. (2007) argue, problemsheds are a socio-political construct and are thus more appropriate as a unit of management, and hence regulatory organisation, than a watershed or river basin. Watersheds, in any case, frequently influence – and are influenced by – factors outside their hydrological boundaries (Islam & Susskind 2013, pp. 51–52). As Islam and Susskind point out, the source of many water management difficulties stems from the fragmented or bounded view of water as a 'natural object', or at other times, as a 'societal issue' or at still other moments as a 'political construct'. They argue that the components of any water governance puzzle can fit together in so many different ways that it is practically impossible to use 'reductionist' or traditional 'systems engineering' methodologies to resolve complex water management problems (Islam & Susskind 2013, p. 42). The concept of a problemshed takes account of these complexities.

The Ganges–Brahmaputra problemshed is defined here as those parts of the Ganges–Brahmaputra–Meghna hydrological basin that fall within the political boundaries of Nepal, Bhutan, Bangladesh and northern India. This basin is shown in Figure 1. This river basin extends to the Tibetan plateau, however the problemshed of interest here does not include China because the centre of gravity for foreign-led interventions in transboundary water governance lies in the interaction between India, Nepal, Bhutan and Bangladesh. Although the main foreign-led interventions – the Abu Dhabi Dialogue and the South Asia Water Initiative – do include China (and Pakistan), these are in effect treated as separate projects from those of improving interactions between Nepal–India–Bangladesh in the Ganges basin, and Bhutan–India–Bangladesh in the Brahmaputra basin.

That is not to belittle the significance of China, Chinese interests in the water resources of Nepal, Bhutan, India and Bangladesh, and the relationships of those countries with China. On the contrary, as Earle et al. (2015, pp. 41–42) point out, '[t]he advent of China in the [Ganges–Brahmaputra–Meghna] basin's hydropolitics, like in the Mekong and the Nile, has brought further complications and uncertainties'. Of those further complications and uncertainties brought on by the advent of China, only those of interest to the riparian states are of interest here.

China is relevant here only insofar as India, Nepal, Bhutan and Bangladesh perceive, engage or otherwise relate to China as part of the hydropolitics among one another. For instance, it is relevant here that India evokes the spectre of China to securitise water in the problemshed. As Hill explains, the activities of

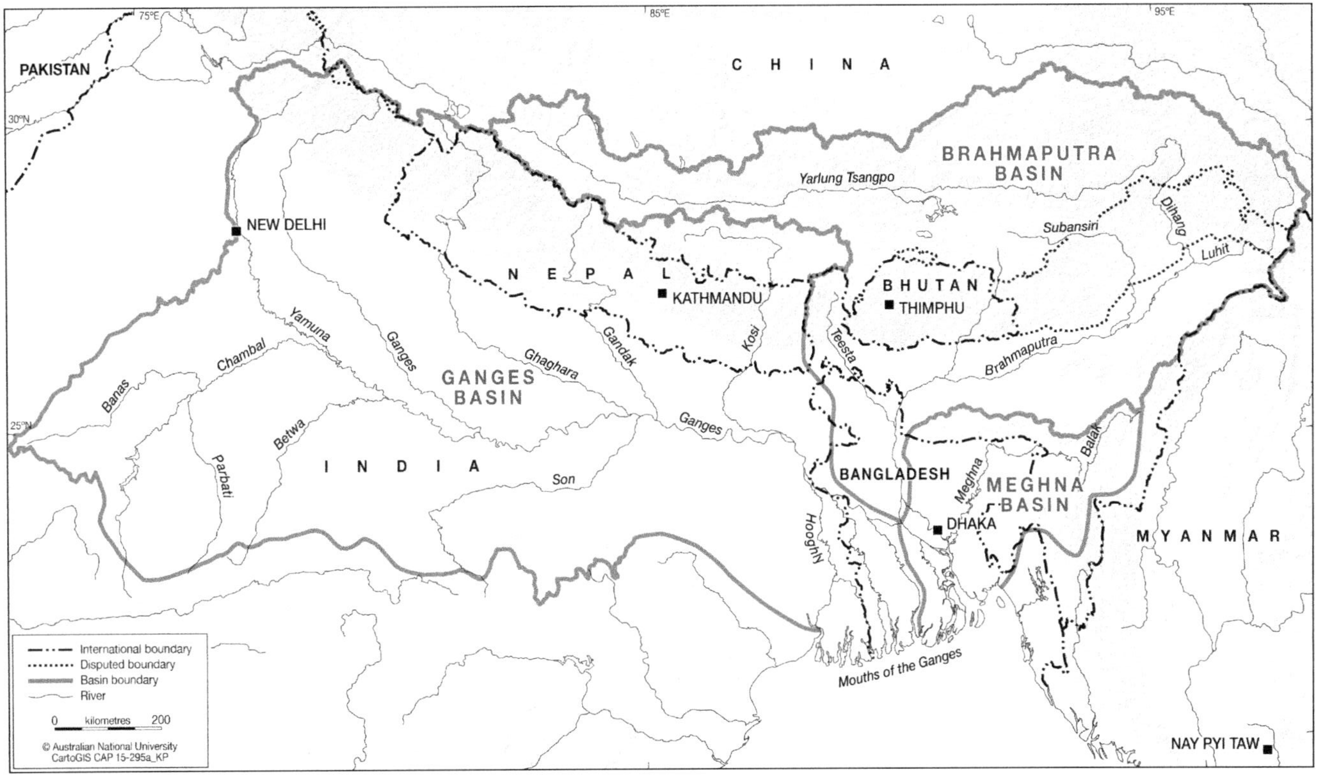

Figure 1 Map of the Ganges–Brahmaputra–Meghna basin

Source: © The Australian National University, College of Asia and the Pacific, CartoGIS.

Chinese hydropower companies in Tibet have led to a great deal of commentary in the Indian media and beyond. Many in the region and beyond are, according to Hill (2015, p. 734), beginning to question whether a dispute between India and China over the utilisation of the waters of the Brahmaputra may spur confrontation and eventually escalate into outright conflict.

Of specific concern is how the damming of the Brahmaputra and its tributaries will affect the flows of water into northeast India, especially during the dry season. For the purpose of my argument, however, the exact details of China's plans on the so-called Great Bend of the Brahmaputra River are less relevant than the perceptions within India and Bangladesh, and how those perceptions affect hydropolitics with China.

That is not to say that China is not significant in the region, or that it does not face water governance challenges. It does, and these are of interest to international organisations and foreign aid donors too. But these Chinese water challenges do not form part of the Ganges–Brahmaputra problemshed because, as per the definition, a problemshed is a unit of feasible water management. The part of the Brahmaputra River which lies within China (and is there called the Yarlung Tsangpo) cannot readily be managed under the same water governance regime as the part of the Ganges–Brahmaputra–Meghna basin on the south side of the Himalayan range.

Transboundary water governance, cooperation, conflict and interaction

Water cooperation refers to a generally agreeable state of water governance, which, according to Dore (2014, pp. 195–197), is:

> the multi-layered interplay of water-related negotiations, agenda-setting, preference shaping, decision-making, management and administration between many actors (including organisations) in the state-society complex, at and between different levels and scales, vying for authority or influence, and constrained or enabled by evolving norms and institutions.

In other words, water governance is the whole complex system of stakeholders, events and agendas at all levels. A more workable understanding is that '[w]ater governance refers to the range of political, social, economic and administrative systems that are in place to develop and manage water resources, and the delivery of water services, at different levels of society' (Rogers & Hall 2003, p. 7). This Global Water Partnership (GWP) definition from 2002 is one of the most commonly used, and although it is not perfect (Araral and Wang 2013, p. 3947, for example, argue that it 'lacks a coherent analytic framework and diagnostic value as it is mainly descriptive'), it is relevant here because it implies that water governance is a complex system. Water cooperation, then, is an emergent property of that system.

Emergence is one of the principles of complexity (Mitleton-Kelly cited in Islam & Susskind, 2013, p. 65). It means that a quality or characteristic is only present through the combination of particular factors or conditions, but is not inherent to those factors in isolation. In other words, 'characteristics of larger units are not simple combinations of attributes of smaller units, but can show new, collective behaviours' (Islam & Susskind 2013, p. 73). Consciousness, for example, is an emergent property of neural processes in the brain. Therefore, if water governance is a 'range of political, social, economic and administrative systems' (Rogers & Hall 2003, p. 7) then cooperation may emerge when those systems develop particular attributes.

Transboundary water cooperation by definition emerges across boundaries. Those boundaries here refer to political boundaries, that is, borders between states. For Grey, Sadoff and Connors (2009, p. 19) the state is the central actor in transboundary cooperation: 'Effective cooperation on an international watercourse is any action or set of actions by riparian states that leads to enhanced management or development of the watercourse to their mutual satisfaction.' This definition, however, is problematic for two reasons. First, it may lead to the misconception that a treaty or international agreement is automatically a measure of cooperation. Second, it presumes governments are the only actors in water governance systems.

A treaty between two or more riparians is not necessarily a measure of cooperation between them. As Warner and Zawahri (2012, p. 215) argue:

> even when international water agreements are signed, it does not mean contracting states are actually cooperating, and the lack of agreement does not mean riparian states are fighting. In other words, the presence of a treaty does not automatically translate into behaviour altering cooperation.

Zeitoun and Jägerskog (2009, pp. 9–13) add that a treaty does not reflect the cooperation (or lack thereof) between states. Some treaties, they argue, stand for so long that their terms become meaningless in the evolving (or devolving) reality of the problemshed. As an example of this they point to the Mahakali Treaty between India and Nepal, which does not hide the lack of coordination between the two states. Its presence was nowhere felt when the 2008 Kosi flood caused widespread destruction on both sides of the border. Zeitoun and Jägerskog (2009, pp. 9–13) also argue that an inequitable treaty may in fact be part of the problem between riparians, and 'the existence of an unfair and ineffective treaty can prevent a more equitable arrangement from being established'.

Moreover, while states are the primary and most powerful actors in transboundary water governance systems, they are not the only ones. As Mollinga (2008b, p. 12) argues, the political contestation of water policies takes place within state apparatuses, but also in the interaction of state institutions with the groups directly and indirectly affected by these policies. In the context of international development, the interaction with foreign aid agencies also affects these policies. Together, this messy process is known as water interaction. When it occurs between states in regard to shared freshwater resources, it is called transboundary water interaction.

Mirumachi defines transboundary water interaction as 'relations of co-existing cooperation and conflicts among communities, groups or states over international or sub-national water, with a focus here on inter-state interaction' (Zeitoun & Mirumachi 2008, p. 299). Central to the idea of water interaction is the understanding that conflict and cooperation can coexist. Strictly speaking, then, cooperation is transboundary water governance marked by low conflict, high cooperation – or, in Mirumachi's terminology, positive interaction (as distinct from negative or neutral interaction). Positive interaction is 'inter-state interaction that generally tends to meet the interests of the actors, and contributes to improvement or sustained relations at the broader political level' (Zeitoun & Mirumachi 2008, pp. 309–310). The term transboundary water cooperation, or just cooperation, is used within this book as shorthand to denote positive interaction, or low-conflict, high-cooperation interaction.

The quality of transboundary water interaction shifts over time in accordance with changes in the socio-political context and specific events. This book examines the effect that international organisations and foreign aid donors have on the socio-political context of the Ganges–Brahmaputra problemshed, and questions whether this effect is sufficient to turn transboundary water interaction between India, Nepal, Bhutan and Bangladesh into 'positive interaction'.

Track II dialogue

One mode of water governance is that of deliberative governance. Deliberative water governance is the 'constructive engagement in water governance arenas through promotion of inclusive, deliberative processes that emphasise different perspectives, critical analysis, learning and institution-building whilst respecting rights, accounting for risks, acknowledging responsibilities and fairly distributing rewards' (Dore 2014, p. 195). Within deliberative water governance there are, as Dore (2007, pp. 205–234) has shown, multiple tracks. Each track can be thought of as a form of dialogue or diplomacy.

Track I Diplomacy, according to Petersen-Perlman and Wolf (2015, p. 5), is the bringing together of riparian states (typically through mediation and facilitation) for the purpose of concluding a formal agreement. In contrast, Track II Diplomacy aims to develop strategies on the ground by engaging state and non-state actors informal dialogues, research and studies, and capacity building. Track III Diplomacy, they continue, addresses policies that affect local-level water governance. Of course none of these strategies are mutually exclusive; for example, Track II Diplomacy efforts may eventually lead to the initiation of more formal, Track I discussions (Petersen-Perlman & Wolf 2015, p. 5). Dore (2014, p. 203) agrees that Track II dialogues often emerge as complementary or parallel forums to Track I dialogues, and thus enhance the effectiveness of state-centric processes.

Track II dialogue is also a form of dispute resolution, and can be a complement to formal negotiations over a contentious issue or set of issues. Because Track II dialogue is most often applied in situations of conflict or dispute, a third party is likely to be involved. Jones (2015, p. 24) notes that Track II dialogue is more often than not facilitated by a third party. He defines the process as:

> unofficial dialogues, generally between two antagonistic parties, and often facilitated by an impartial Third Party and involving individuals with some close connections to their respective official communities, focused on cooperative efforts to explore new ways to resolve differences over, or discuss new approaches to, policy-relevant issues.

Salame (2014, p. 135) shows that Track II initiatives are most useful when all parties recognise that there is a divergence of views while neither side wants the situation spiralling out of control – especially on vital issues such as transboundary water governance. She argues that Track II initiatives build cooperation slowly and steadily, and allow decision-makers to engage in political dialogue based on trust and mutual understanding.

Track II dialogue can certainly bring state and non-state actors together in an informal setting to discuss issues of mutual concern; however, as I argue, this method alone cannot build transboundary water cooperation. Furthermore, failure of Track II dialogue may undermine state efficacy in improving water governance. As Dore (2014, p. 203) shows in the case of the Mekong basin, loss of faith in Tracks I and II has led to the emergence of increasingly influential Tracks III and IV – civil society to civil society interaction – that do not privilege state involvement and may proceed without it.

Of all the tracks, Track II Diplomacy or dialogue is of most relevance here as it is the primary method through which international organisations and foreign aid donors, especially the World Bank-led South Asia Water Initiative, are attempting to increase transboundary water cooperation in the Ganges–Brahmaputra problemshed.

1 International actors in the resolution of transboundary water conflicts

This chapter surveys the literature on the resolution of international disputes, and the role of third parties in negotiation. The first section of this chapter provides an overview of knowledge about the role of third parties in conflict resolution and negotiations. Traditionally those roles have been mediation, conciliation, arbitration and adjudication of disputes. The emphasis was on reaching agreement over specific issues within the limited scope of the negotiation process. The effects of external factors, such as culture or power dynamics between parties to the conflict, were rarely taken into account.

In the late 1960s, however, a new method of international negotiations emerged: Track II Diplomacy or dialogue. This is the idea that the relationships built and ideas discussed through informal channels or 'pre-negotiation' may have an overall positive effect on levels of cooperation between conflicted parties. The role of third parties in Track II dialogue is that of facilitation. The focus shifted away from reaching consensus and towards improving relations between stakeholders so that it would be easier to negotiate agreements in the future.

The discussion then turns to transboundary water governance specifically. Very little has been written about how international organisations and foreign aid donors could contribute to transboundary water cooperation. This is despite the fact that the involvement of aid donors and international organisations in transboundary water governance is increasing the world over.

What little literature does exist on the role of third parties in increasing transboundary water cooperation is covered in the second section of this chapter. Some of this literature suggests that the most important role for external actors is in facilitating benefit sharing arrangements between riparians. This, however, confuses cause and effect. Benefit sharing is largely an outcome of positive water interaction rather than a driver of it. Thus, while neutral third parties may be well positioned to mediate benefit sharing agreements (in the same way that a neutral third party can be used to mediate any negotiation), it is unlikely that riparians will come to the formal negotiation table without some degree of trust between them and political will to resolve the issues at hand. In other words, transboundary water cooperation is a precondition of benefit sharing rather than *vice versa*.

The main focus in the literature on third party interventions in transboundary water governance is on dispute resolution rather than on creating positive water interaction. The difference here is subtle but important. Dispute resolution is about coming to terms on a specific issue, whereas positive interaction is a quality of governance or a situation within which disputes may exist, but which is overall cooperative. Only in recent years has the negotiation literature begun to take into account the broader, contextual factors that may sustain water disputes – much as the scholarly thinking on negotiation and diplomacy first centred on dispute resolution, and only slowly began to consider the effects of contextual factors such as culture and power dynamics.

The emerging thinking is that international actors may have a role in facilitating general processes (similar to, or even explicitly called, Track II dialogue) that could sustain cooperation. This is sometimes known as water- or hydrodiplomacy; a concept also discussed in this chapter. Understanding this function helps make sense of what the World Bank and other aid donors are doing in the Ganges–Brahmaputra problemshed.

The resolution of international conflict through third parties

The role of third parties in conflict and cooperation is well documented in the literature on negotiations. This body of literature is extensive and negotiation theories are plentiful. However, only negotiations relating to disputes over environmental resource management are of interest here. Likewise, in transboundary water interaction – that is to stay, interaction primarily between states – it is state-to-state negotiations that are of most relevance. As Elver (2006, p. 892) points out, '[t]here is no question that international water disputes generally depend on state-to-state negotiation'.

State-to-state negotiations can be separated into two categories. The first type of negotiation is that of formal, bounded conflict usually over a single issue. The second type of negotiation is Track II dialogue, or a loose, complex, informal set of overlapping processes usually over numerous issues and usually between government and non-government actors. Importantly, a negotiation of either kind is neither the beginning nor the end of transboundary water interaction. As Tiwary (2006, pp. 1686–1687) states, '[t]he politics (national or domestic) over water do not abruptly end with an international treaty; it just enters a new phase of inter- and intra-nation level scrutiny and criticisms'.

This section first outlines the traditional roles of third parties in negotiations, then discusses the role of power dynamics between parties to a dispute, and the cultural aspects of negotiation. The third sub-section focuses on Track II dialogue and the role of the third party as facilitator.

Negotiation and dispute resolution: the third party as mediator, conciliator, arbitrator or adjudicator

Formal negotiations can be seen as a way of containing conflict within mutually acceptable and non-violent processes. The idea of negotiation as bounded conflict is, for Dore (2014, p. 200), 'critical for complex water governance where there is, more often than not, a divergence of interests between actors who also recognise there is a degree of mutual interdependence in resolving problems'. The divergence of interests is the root of conflicts, and formal negotiation is also primarily about conflict resolution.

According to Cosgrove (2003, p. 37), '[n]egotiation is the means of dispute resolution most often employed by states when trying to resolve any international conflict, including those over transboundary water resources'. But different states will approach a formal negotiation differently at different times and for different purposes. It is possible to discern a national negotiating style, but this is fluid and comprises several difficult to quantify elements: culture, history, political systems and position in the world (Faure & Sjöstedt 1993, p. 5). Negotiations are complex and most negotiation experts suggest that 'strong and effective agreements are more difficult to reach as the number of actors increases' (National Research Council of the National Academies 2012, p. 85).

Formal negotiations can take various forms depending on the issues at stake and the number of states involved (e.g., bilateral talks, diplomatic correspondence, international conferences) and these can be used at all stages of the conflict, although:

> [n]egotiations are considered merely as the first step that states usually take in resolving their dispute. If they fail or if parties are unable to enter into negotiations altogether, other means of dispute settlement are available to them, and all are based on the involvement of a neutral third party.
>
> (Cosgrove 2003, p. 37)

The involvement of a third party is more often than not a consideration in formal negotiations, where parties to the conflict may seek assistance in reaching agreement. That is because, as Faure and Sjöstedt (1993, p. 7) explain, the purpose of a formal negotiation is 'to find a formula for the distribution of a contested value or set of values between the negotiating parties. Thus negotiation is a joint decision-making process through which negotiating parties accommodate their conflicting interests into a mutually acceptable settlement.'

Third parties, then, can be neutral shepherds of a negotiated settlement. The roles that third parties can play in a formal negotiation are relatively clear: mediation, conciliation, arbitration, adjudication. A neutral third party (be it a state, joint body, international organisation or an individual) can offer 'good offices' to the conflicting states, and act as a 'go-between' in order to persuade them to enter into negotiations (Cosgrove 2003, p. 38). Mediation is a step towards more active third-party participation in the formal negotiations, because a mediator provides assistance to the disputing parties in finding a solution (Cosgrove 2003, p. 38). Conciliation is an effort to reconcile opposing contentions; the parties to the dispute are left free to accept or reject the proposals formulated by third parties (Cosgrove 2003, p. 40). However, as Cosgrove (2003, p. 39) points out, '[t]he boundaries between good offices, mediation, and conciliation are sometimes blurred, and one procedure can often lead to another'.

Two further roles for third parties include arbitration and adjudication. Arbitration allows parties to the conflict to nominate the arbitrator(s) who will hear the dispute, and to determine the law governing the substance and procedure of the case, where the dispute will be heard, and the language of the arbitration (Cosgrove 2003, p. 41). States embroiled in transboundary water conflict may also agree to have their cases heard by an *ad hoc* or permanent court, such as the International Court of Justice (Cosgrove 2003, p. 42). However, as Islam and Susskind (2013, p. 133) point out, assigning a higher authority has become an increasingly unpopular approach to water conflict resolution in recent years and instead a negotiated settlement is considered preferable. Finally, a third party could conduct an inquiry or fact-finding process designed to produce an impartial finding of disputed facts because, as Cosgrove (2003, p. 39) shows, many international disputes arise from disagreements over questions of fact.

Nonetheless, reaching a settlement or formal agreement through mediation, conciliation, arbitration or adjudication is not necessarily the same as developing cooperation between conflicted parties. As Qaddumi (2008, p. 8) points out, '[u]ltimately, cooperation rests not on objective measure of gains to be had, but rather on the subjective perceptions held by these various groups and how these are played out in the policies, institutional arrangements and, finally, treaties'. It is easy but misleading, according to Habeeb (1988, p. 27), to think of negotiations as simply a series of offers, demands and concessions, in which the two sides inch toward an agreement. Yet this is the basis of the benefit sharing approach to creating transboundary water cooperation. Benefit sharing is described in the final section of this chapter. Instead, Habeeb (1988, p. 13) argues that the context of culture and power dynamics between actors has as much of an influence on building cooperation as does the process of formal negotiations. The importance of the cultural factors is largely overlooked in the literature on negotiation theory except that, as Zartman (1993, p. 17) claims, 'the only negotiations outcome that culture is purported to explain is failure'.

Context: the role of culture and power in influencing cooperation

In addition to the well-defined roles of mediation, conciliation, arbitration and adjudication, third parties may be able to assist riparians in pre-negotiation efforts. This role could include ensuring that the appropriate stakeholders are at the table, that they are well prepared, and specify a timetable, ground rules, agenda and fact-finding procedures before formal negotiations begin. Islam and Susskind (2013, pp. 134–135) argue that it is these pre-negotiation efforts, rather than what happens within the formal negotiation process, that largely determine the success or failure of attempts at transboundary water dispute resolution. Faure and Sjöstedt (1993, p. 12) add that '[o]utcome is the function of all of the other dimensions of negotiation: actors, structure, strategy, and process. Accordingly, the impact of culture on these other elements of negotiation eventually also indirectly influences the outcome'.

Culture, then, affects how riparians approach negotiation processes, but it also affects the role and approach of third parties. As Faure and Sjöstedt (1993, p. 6) point out, any actor in a negotiation, be it a party to the dispute or a mediator, is likely to assess foreign cultures through their own cultural lenses. Those characterised by ethnocentrism tend to interpret and judge other cultures by their own standards: '[t]he effect may be an inclination to belittle or misinterpret the intentions or design of other parties involved in the negotiations. To the extent that negotiators behave along these lines, they are likely to impede an agreement.' This tendency limits the neutrality of any third party involved in transboundary water negotiations.

Similarly, Tiwary (2006, pp. 1686–1687) shows that considerations outside the negotiation table affect the outcome of formal negotiations. He argues that above all else, 'international water conflict resolution requires considerable political will on all the sides' and sustained motivation to reach agreement. Tiwary (2006, pp. 1686–1687) also argues that states negotiate not only in reaction to other riparians, but also in reaction to domestic politics. For example, in nations where threats of water security are more apparent, or in a relatively weak state in a negotiation, there is a risk that negotiators will be labelled as villains who compromised over national well-being. Fear of this colours the negotiation dynamic. Thus, understanding the cultural context and power dynamics between riparians is important in understanding informal negotiations over transboundary waters, and political will on the part of the stakeholders is an important factor in conflict resolution.

Similarly, power relations between stakeholders in a negotiation also affect its outcome. Habeeb (1988, p. 129) identifies three types of power: aggregate structural power, issue power, and tactical or behavioural power. The first is an actor's total resources and capabilities (for example, India has the greatest political, economic, military and soft power strength in the Ganges–Brahmaputra problemshed). Issue power is an actor's resources and capabilities within the context of a specific issue (for example, a relatively small and weak country may have great issue power if, like Bhutan, it is upstream of a hydro-hegemon that, like India, has a strong interest in its freshwater resources). The third type of power is tactical, or behavioural power, and refers to an actor's ability to use its power resources to attain objectives. (For example, Nepal has low tactical power as it is unable to harness its political resources to obtain water-related objectives due to numerous and conflicting other priorities on the political agenda.)

It is issue power, argues Habeeb (1988, p. 129), that largely determines the process and outcome of state-to-state negotiation. Furthermore, the aim of tactical power – an actor's ability to use its power resources – is 'to alter the issue power balance in a more favourable direction'. Habeeb (1988, pp. 136–137) shows that negotiation processes are in essence efforts by each side to change the pre-negotiation issue power balance in a direction more favourable to their interests. The final agreement thus reflects the final issue power balance, but says nothing about the aggregate power balance between the two sides. It only redefines the two sides' interdependent relationship in a particular issue area.

The case of transboundary water governance in the Ganges–Brahmaputra problemshed is illustrative of this dynamic. Nepal, Bhutan and Bangladesh are trying to use tactical power to negotiate more favourable water outcomes with India. Only Bhutan has succeeded in increasing issue power (this is discussed further in Chapter 3). But none of it affects the overall power asymmetry between the hydro-hegemon and the weaker riparians.

Pohl et al. (2014, p. 17) acknowledge that '[b]asin politics are often compounded by power asymmetries, begging the question of how to deal with riparian hegemons'. They argue that international actors must engage with basin hegemons (Pohl et al. 2014, pp. 25–27). This is in line with Zeitoun and Jägerskog's argument from 2009 (p. 12) that one way to address power asymmetry is to 'create conditions to encourage basin bullies to transform into basin leaders'. The effects of power asymmetry between conflicted parties may to some extent and in certain circumstances be attenuated by the involvement of a neutral third party in negotiations. One method for involving a neutral third party is Track II dialogue. This is discussed next.

Track II dialogue: the third party as facilitator

Track II dialogue has its genesis in the mid-1960s when the prevailing realist notions of international negotiations began to be challenged. The view emerged that human factors such as dialogue, communication, values and relationships between individuals could be equally important in avoiding and resolving international conflicts as state power. From the very beginning, points out Jones (2015, pp. 13–14), Track II dialogue, like conflict resolution generally, 'has struggled to establish credibility in the eyes of hard-nosed, realist-oriented officials and academics'.

The new process 'controlled communication', and it consisted of informal, unofficial workshops, chaired by a neutral third party who facilitated the protagonists' mutual analysis of problems with the aim of helping them develop solutions that were not apparent through traditional diplomatic techniques. This method attracted the attention of the academic community and soon a small but active 'scholar-practitioner' community arose, intent on furthering these new ideas (Jones 2015, pp. 13–14). This method fits in well with other new ideas about democracy, such as deliberative governance. Deliberative governance, like Track II dialogue, is about creating and nurturing the 'preconditions for positive changes and their institutionalisation by strategic nudging within these admittedly complex arenas' (Dore 2014, p. 199).

The role of the third party in this nascent 'interactive problem solving' approach was further developed by Herbert Kelman, a Harvard-based social psychologist. Kelman (cited in Jones 2015, pp. 14–15) defined his model as 'an academically based, unofficial third party approach, bringing together representatives of parties in conflict for direct communication'. He saw the role of the third party in this arrangement as being different to that of a traditional mediator. Mediators propose solutions (and arbitrators impose them), while third parties in Track II dialogue facilitate a process whereby solutions will emerge out of the interaction between the parties themselves. The main tasks of the third party in Track II dialogue are to provide the setting, create the atmosphere, establish the norms and offer the occasional interventions that make it possible for such a process to evolve, writes Jones (2015, pp. 14–15). This is the model on which third party interventions in the Ganges–Brahmaputra problemshed are also based.

Jones (2015, p. 87) sees the involvement of a third party as the defining feature of Track II dialogue: 'Not all Track Two requires an outside third party. But the majority of Track Two processes, and the great majority of conflict resolution-oriented Track Two processes, feature a third party.' But is a third party really necessary in Track II dialogue? Could people from societies in conflict quietly come together themselves to brainstorm about possible ways forward?

Most of the literature on Track II dialogue takes the view that a third party is necessary, and Jones (2015, p. 88) argues that the problem at the heart of Track II dialogue is the inability or unwillingness of the protagonists to the conflict to come together of their own accord. Furthermore, he argues, brainstorming by the parties themselves may not be enough, and under the wrong circumstances it may do more harm than good. Jones argues that Track II dialogue is not just about conflicted parties engaging in brainstorming. Rather, it is conflict parties engaging in a particular kind of *facilitated* discussion. Absent this role, he continues, the conversation can degenerate into a frustrating bargaining session which holds little promise. This is especially so in a highly charged, emotional situation where the conflict is deemed to be intractable. In such circumstances, asks Jones (2015, p. 88), what is there to bargain over?

Yet, third parties, though central to Track II dialogue, may also contribute to the worsening of a situation. As John Burton, the Australian former diplomat and early adopter of Track II dialogue, has said, 'the intervention of a third party into relationships between others is a delicate task and can easily do more harm than good' (Burton cited in Jones 2015, p. 91).

How, then, should the third party behave to minimise harm and maximise good? The third party should have some knowledge of the specifics of the conflict, and a firm grasp of group dynamics is essential (Jones 2015, p. 99). Neutrality, however, may not always be necessary. Jones argues that so-called insider-partial third parties may be just as effective. The most salient characteristic of the third party for Jones (2015, p. 104) is that they be able to run the process in a fair and impartial way and will not, due to their own views, disadvantage anyone at the table.

Third parties will always act out of a mixture of motivations and objectives (Chapter 4 examines these drivers of external actors in the Ganges–Brahmaputra basin). This does not necessarily impair their neutrality or effectiveness, but '[w]hatever the third party's motivations are, it is vital to reflect on them and be honest about them up front' (Jones 2015, p. 103).

What other qualities must the third party possess in order to fulfil the role of facilitator? Saunders writes of the 'wise citizens, well prepared'. For Jones (2015, p. 95), this definition illuminates two important characteristics of the third party.

First, political sophistication, and second, the ability to make people in conflict gradually feel safe enough to open themselves up to the third party, and to each other.

One of the tasks of the third party in Track II dialogue is to build a relationship of trust between the parties in the conflict to a level that is sufficient for them to be drawn into a conversation in which they will gradually reconsider the meaning of the conflict and simultaneously reveal their innermost perceptions of the conflict to each other. Moreover, two parties who profoundly mistrust each other must be able to trust someone in the room until they can come to feel at least some trust for the other side. This is the role of the third party (Jones 2015, p. 102).

To gain and retain the trust of the parties in the dispute is thus, according to Jones (2015, p. 101), the single most important factor in the eventual success of conflict resolution. The environment within which the third party begins working is, by definition, particularly lacking in trust, and once broken, trust is not easily regained.

In conclusion, trust between parties is a necessary condition for Track II dialogue to succeed in contributing to the resolution of international disputes and building cooperative relationships between states. At the same time, Jones (2015, p. 140) points out that 'real peace is made as a result of a complex and interlocking web of factors, and that Track Two, however important in getting a dialogue going, is but one of these'. The following section of this chapter turns to the idea of what Jones' 'real pace' looks like in regard to transboundary water interaction.

Towards positive water interaction through third party interventions

The year 2013 was the United Nation's International Year of Water Cooperation (UN 2013), and saw the publication of books such as UNESCO's scholarly compilation, *Free Flow – Reaching Water Security Through Cooperation* (UNESCO 2013). This reflected and cemented a growing enthusiasm within and outside academia for cooperation, collective action, good governance and so on, as a means of tackling serious challenges facing the globe: climate change; water scarcity; communicable disease; volatility in international food, fuel and financial markets; and the depletion of common-pool resources, such as fisheries (Evans et al. 2015, p. 2). Stemming from this idea of cooperation, better water governance was identified as 'absolutely the top priority' for human security (Jägerskog et al. 2014, p. viii). By 2010 '[m]any studies at international level have indeed identified the water crisis not as one of scarcity but as one of governance' (Jägerskog et al. 2014, p. xi). But while good governance and transboundary cooperation have been generally agreed upon as a goal (a topic of further discussion in Chapter 5), few scholars are asking how it can be achieved, let alone how it can be achieved with the assistance of international organisations and foreign aid donors.

In 2009, Jägerskog and Zeitoun began asking these questions in a report for the Stockholm International Water Institute called *Getting Transboundary Water*

Right: Theory and Practice for Effective Cooperation. The report asks 'What is the quality of cooperation? What does "cooperation" mean in the first place? Why and under what conditions do states cooperate?' (Jägerskog & Zeitoun 2009, p. 6). But the answers it provides are ambivalent at best. In the report's second of five essays, Zeitoun reiterates his theory of hydro-hegemony. He and Jägerskog answer the three central questions of the report by arguing that states cooperate once power asymmetry between them is addressed (either by being influenced or challenged) and they also list general ways in which third parties can assist riparians in influencing or challenging power asymmetry (Zeitoun & Jägerskog 2009, pp. 9–13).

This is a useful way of thinking about water interaction, but it does not say much about the conditions necessary or sufficient for transboundary cooperation to emerge. After all, as Zeitoun acknowledges, transboundary cooperation – and conflict – can flourish irrespective of the presence of a hydro-hegemon (Zeitoun & Allan 2008, p. 10). This idea is echoed more than five years later in the report *The Rise of Hydro-Diplomacy*, which states that '[h]egemons are not necessarily impediments to cooperation; they can also facilitate it' (Pohl et al. 2014, p. 27). In other words, the presence or absence of a basin hegemon is not a necessary condition for transboundary cooperation, although hegemonic or counter-hegemonic interactions do affect political will, trust between riparians and facilitated dialogue between stakeholders.

The remainder of *Getting Transboundary Water Right* focuses on frameworks for reaching agreement between riparians on sharing the benefits of a transboundary basin. Granit and Claassen 'present the Transboundary Water Opportunity (TWO) Analysis, which provides stakeholders with a framework to identify a variety of opportunities for a basin' (Jägerskog & Zeitoun 2009, p. 7) while Grey, Sadoff and Connors (2009, pp. 15–16) argue that states 'cooperate when the net benefits of cooperation are perceived to be greater than the net benefits of non-cooperation, and when the distribution of these net benefits is perceived to be fair'. In both of these essays the equitable sharing of benefits is understood to be the cause of transboundary cooperation, rather than the effect of it. The role of third parties is to mediate as riparians negotiate the optimal spread of benefits for all stakeholders. This is a classic role of the third party in formal negotiations, which do not take into account contextual factors or the relationship between parties to the conflict.

Benefit sharing: cause or consequence of cooperation?

Sadoff and Grey have been two prominent contributors to the embryonic literature on – and the practice of – transboundary water cooperation through third party interventions. Since 2002 they have claimed that transboundary water is largely a matter of optimising benefits, and therefore the role of third parties is that of a neutral broker mediating the development of a constellation of possible win–win outcomes (Sadoff & Grey 2002, pp. 389–403). Grey, Sadoff and Connors (2009, pp. 15–16) argue that states do not cooperate because of any ethical imperative, but rather because it can be pragmatic to do so: 'states work together when doing so offers special economic and political advantages over unilateral development, and when these larger benefits are shared.'

In the benefit sharing approach, the role of the third party is to nudge riparians away from unilateral national agendas and toward a shared cooperative agenda, one which provides 'benefits that exceed the sum of the two non-cooperative national agendas, and will thus have become the rational choice of each sovereign nation' (Sadoff & Grey 2005, pp. 420–421). Cooperation, then, is a rational choice, and the natural product of maximised outcomes. For Grey, Sadoff and Connors (2009, p. 16), all that stands in the way of creating transboundary cooperation is the puzzle of how best to maximise for all riparians four types of benefits:

1. environmental benefits to the 'river' (e.g. improved water quality, conserved biodiversity);
2. economic benefits from the 'river' (e.g. increased food and energy production);
3. reduction of costs because of the 'river' (e.g. reduced geo-political tensions, enhanced flood management); and
4. benefits beyond the 'river' (catalysing wider cooperation and economic integration).

Moreover, they claim that '[t]he broader the basket of benefits, the greater is the scope for structuring mutually beneficial cooperation' (Grey, Sadoff & Connors 2009, p. 16).

Of particular significance here, is the fact that Sadoff and Grey (and Connors) have all been involved in various capacities with the South Asia Water Initiative (SAWI) and the Ganges Strategic Basin Assessment (GSBA). Indeed, the guiding principle of the Ganges Strategic Basin Assessment was benefit sharing. The purpose of this World Bank study was to gather comprehensive hydrological, economic and socio-political data from which draw informed conclusions about maximising benefits to the river, from the river, of the river and beyond the river. Yet, as Chapter 7 illustrates, the laying out of possibilities for positive-sum outcomes based on the hydrological, economic and social modelling of the Ganges basin did not bring riparians rushing to the negotiation table to capitalise on these potential benefits. The Ganges Strategic Basin Assessment was rejected by the riparians and failed to establish a new era of transboundary cooperation based on mutual benefit sharing. Why? Much like the idea of cooperation through benefit sharing, the Ganges Strategic Basin Assessment fails to account for the broader socio-political context in which the basin is situated. As Habeeb (1988, p. 27) has argued, it is easy but misleading to think of conflict resolution as simply a series of offers, demands and concessions, in which the two sides inch toward an agreement.

The major weakness of seeing benefit sharing as the sufficient condition of cooperation is that it does not account for all the other influences on states engaging with each other (or not) over shared water issues. For example, power asymmetry between riparians, underlying historical grievances, and non-water related interests and agendas are not explicitly taken into consideration. Sadoff and Grey (2005, p. 420) do acknowledge that '[a]chieving international cooperation is always a long and complex journey, for which there is no single path and few short cuts'. They also note that the optimal mode of cooperation will be different in every basin depending on hydrologic characteristics, the economics of cooperative investments, numbers and the relationships of riparians, and the costs of parties coming together.

Perlman-Petersen and Wolf's study (2015, p. 2) of transboundary water cooperation also finds that there is no one way to initiate cooperation because each transboundary basin is unique. They argue that '[i]ndividual policy makers in a nation making decisions about cooperation operate within the historical context of their nations, fed by a set of external and internal drivers of decision making ... They do not consider benefits alone.' Global trends, for example, can 'also exert influence on cooperation, given the history of ideas and experience regarding international waters that nations and their partners contemplating cooperation can draw from' (Perlman-Petersen & Wolf 2015, p. 4). Because of this, Perlman-Petersen and Wolf (2015, p. 4) argue that '[i]t is necessary both to address past and present grievances as a prerequisite for market-driven solutions' for enhancing transboundary water cooperation.

Sadoff and Grey's definition of benefits as going 'beyond the river' also hints at the importance of non-water related interests affecting water interaction. But they do not address how power asymmetry between riparians could stand in the way of benefits being shared. Moreover, the benefit sharing approach does not account for the fact that these or other issues may first need to be resolved before talk of maximising benefits can even begin. Their theory of benefit sharing is one that belongs in the literature of formal negotiations rather than of water conflict and cooperation. As Qaddumi (2008, p. 8) points out, cooperation is not an objective measure of gains to be had, but rather an emerging property of the subjective perceptions held by the stakeholders, and it plays out in the policies, institutional arrangements and, finally, treaties that these stakeholders develop.

Maximising benefits across a river basin is a tactic that has many merits and may be appropriate for cementing cooperative outcomes if, and only if, the necessary conditions for cooperation are met first, that is, political will, trust between riparians, and dialogue among stakeholders. The sharing of benefits may be an outcome of cooperative water interactions, but it cannot be the cause of them. Similarly, Perlman-Petersen and Wolf (2015, p. 2) find that objective or 'paper' benefits are only the starting point: 'benefits are necessary, but they are not sufficient to induce cooperative action.'

Complexity, cooperation and hydro-diplomacy

Like Sadoff and Grey, Islam and Susskind (2013), also advocate the development of non-zero-sum approaches to negotiations, but they see water governance (or water management networks, in their terminology) as a complex adaptive system. They understand that by definition there are no straightforward, predictable and replicable solutions to complex problems. Islam and Susskind (2013, p. 90) identify three types of water management problem – simple, complicated and complex – and argue that only '[s]imple problems are amenable to optimisation because they involve easily identifiable and neatly-bounded elements of water management that respond in predictable ways and pose challenges about which there is almost complete agreement with regards to means and ends'.

In other words, if high-conflict, low-cooperation transboundary water interaction was a simple problem, then the optimisation of benefits could be a satisfactory solution. But water interactions are complex, and thus cannot be 'solved' directly. Instead, Islam and Susskind argue that practical solutions to water conflicts 'are most likely to be found through a negotiated and joint problem-solving approach that blends science, policy, and politics to understand … and manage … complex water problems' (Islam & Susskind 2013, pp. 17–18). For Islam and Susskind (2013, p. 89), water conflicts are most constructively understood as differences in values and how to translate those values into policies and actions within the political domain. As such, the role of the third party is to assist in the translation of those values into equitable actions.

Specifically, the Water Diplomacy framework calls for third parties to act as professional mediators or neutral facilitators in water conflict negotiations, and Islam and Susskind (2013, pp. 144–149) describe in great detail the tasks that this entails. These tasks are very similar to those outlined by Jones (2015) throughout his book *Track Two Diplomacy in Theory and Practice* and described above. Indeed, there are many similarities and complementarities between the emerging concept of hydro-diplomacy and that of Track II dialogue. As Jones (2015, p. 147) points out, 'Track Two often plays its greatest role … in assisting the *process* of developing new ideas and incorporating them into negotiations, rather than in necessarily providing specific *outcomes* to officials [emphasis in original]'. It is also about building trust between conflicted parties and helping them learn about each other so that they can develop a more accurate understanding of the complexities each side must deal with (Jones 2015, p. 147). This is also largely true of hydro-diplomacy.

Track II dialogue is a particularly appropriate method for addressing regional security issues, such as transboundary water interaction, in which there might not necessarily be a specific conflict to be resolved but where an atmosphere of distrust and lack of cooperation nonetheless persists (Jones 2015, p. 26). Transboundary water interactions are also an appropriate context for facilitated dialogue because, as Wolf (1998, p. 10) points out, '[w]ater is, by its nature, an interdisciplinary resource – the attendant disputes can only be resolved through active dialog among disciplines'. The importance of dialogue between different groups of stakeholders as a means of building cooperation is beginning to catch on in both the scholarly literature on water conflicts, and the so-called grey literature of government and private sector output. Perlman-Petersen and Wolf (2015, p. 5), for example, identify Track II Diplomacy as one of the strategies available to third parties in pursuing the development of cooperative water interaction.

In 2014, the Berlin-based consulting firm Adelphi released a report written by Pohl et al. called *The Rise of Hydro Diplomacy*. The authors argue that one of the most important tasks for international actors is that of strengthening diplomacy between riparian states. Pohl et al.'s report is interesting in that it emphasises the need for political will and trust between riparians for cooperation to flourish. The authors argue that '[u]ltimately, strengthening foreign policy for transboundary waters hinges on creating and reinforcing international institutions that can channel political will into coherent action' (Pohl et al. 2014, p. 42). Pohl et al. (2014, p. 35) also claim that trust is critical but that '[t]rust can only be built through longterm transparency in terms of data-sharing and intentions regarding future infrastructure, which in turn requires long-term engagement'. Thus the element of time is brought into the understanding of how transboundary cooperation can develop.

Thus the potential for international actors to significantly and swiftly improve transboundary water cooperation is very limited. As Perlman-Petersen and Wolf (2015, p. 5) point out, it is not possible for third parties to create a conducive, political environment alone. Nonetheless, they argue, international organisations may provide incentives for cooperation both directly and indirectly by:

1. Providing technical competence and examples of best practices;
2. Assisting in negotiation and mediation skills, including the provision of legal and other water experts; and
3. Facilitating investments in transboundary settings.

Finally, there remains a role for foreign aid donors and international organisations in shaping the international community that in turn affects and influences riparians in basin with low transboundary water cooperation. Zeitoun and Jägerskog provide an example of why this is important. At the 2009 World Water Forum, steps to develop legal and fair water sharing principles were resisted by powerful hydro-hegemons who instead pushed for enshrining in international law the principle of 'hydro sovereignty' (which, according to Zeitoun and Jägerskog (2009, pp. 12–13), 'appears to be a reversion to the discredited Harmon Doctrine that had been abandoned over a century ago'). The role of international actors in promoting transboundary water cooperation, then, is not limited to basins characterised by negative or neutral transboundary water interaction. Instead, Zeitoun and Jägerskog (2009, pp. 12–13) urge that '[w]e must keep our eyes open to recognise and confront the reality of power asymmetry, basin bullies and coercion in order to achieve effective cooperation' at all levels and in all spheres. This, they argue, 'makes the role of third-parties all the more crucial'.

As can be seen, the role of third parties is well established in the literature on conflict resolution (where it can be mediation, conciliation, arbitration etc.) and Track II dialogue (where it is limited to facilitation). But it is not well established in the literature on transboundary water conflicts and cooperation. The prevailing paradigm is that third parties can assist in establishing benefit sharing outcomes for riparians, but that is not the same as building cooperation. Only in recent years have scholars and practitioners begun to turn to the problem of third parties, especially aid donors and international organisations, as facilitators of positive water interaction. This is the concept of hydro-diplomacy, which takes the principles of Track II dialogue and applies them to questions of transboundary water security.

Nonetheless, there is no one path to cooperation, and no certainty for the third party that they will resolve conflicts. Numerous factors play into the dynamics within dispute resolution processes and the contexts in which they occur: culture, domestic politics, trust (or lack thereof) between parties, power asymmetries, issue power, the presence or absence of third parties and so on. While all these, and other, factors define the outcome of interactions, dialogues and negotiations, they do not pre-determine them. This is an important insight considering the increasing involvement of international aid donors in transboundary water governance and the growing importance of cooperation in reducing water conflict.

2 Legacies and challenges for water governance in the Ganges–Brahmaputra problemshed

This chapter illustrates that there are numerous issues of concern in the Ganges–Brahmaputra problemshed, but that on the whole they are not a high priority for the riparian states. Improving interaction with co-riparians is not prominent on the political agendas either. As Earle et al. (2015, pp. 41–42) have pointed out, there is a lack of political will in the basin, particularly in India, to support the development of multilateral water governance. This limited political will undermines the efforts of international actors to increase transboundary water cooperation.

The multitude of issues affecting water resource management in the region is both attracting international attention, as well as diffusing it. According to one regional analyst (Anonymous 6, 2014), the biggest transboundary water problem in the Ganges–Brahmaputra problemshed is that there is no problem; or rather, there are so many problems that not one is able to command sufficient attention to resolve it.

Some of these crises arise out of the historical context of the problemshed, particularly the colonial hangovers and the legacy of the Green Revolution. There are also numerous non-water related issues that occupy the Indian, Nepalese and Bangladeshi (but less so the Bhutanese) polities and thus detract from the political will available for dealing with transboundary water issues.

Other crises in the Ganges–Brahmaputra problemshed are global, or challenges faced by policymakers the world over: rapid population growth; rising demand for food, energy and water; unprecedented urbanisation; climate change; and socio-political conflict arising out of resource scarcity. These challenges and legacies are the focus of this chapter.

Legacies from history and ideology

The transboundary water issues in the Ganges–Brahmaputra problemshed today are in part the product of a historical legacy and in part shaped by global factors and concerns. The history of British colonialism established a lasting foundation for water governance practices, while the Green Revolution created unintended challenges.

Part of this legacy is the way in which rivers are understood and managed, namely through the 'command and control' approach. At the same time, river management and transboundary water cooperation are not a top priority for the ruling elites of the riparian states because there are more pressing non-water related considerations. These are examined below.

Colonial hangovers and the legacy of the Green Revolution

The British colonial rule in South Asia has left a legacy for transboundary water governance. For example, water in India has been regulated under different state laws, principles and rules adopted over many decades (Cullet 2007, p. 3), but predominantly Irrigation Acts devised in the colonial era (Siddiqui 2012, pp. 155–156). Irrigation was, according to Hardiman (2008, p. 49), the defining achievement and lasting contribution of British water management in the region. Although pre-colonial South Asia depended on various forms of irrigation, canal systems were not centrally governed and 'water was distributed according to the political power of different groups within local communities of users'. In the pre-colonial era, floods on the Gangetic plain had been understood as a duality of calamity and benefits. Flooding brought nutrient-rich sediment, and rice harvests were traditionally segmented throughout the year to hedge against flood pattern fluctuations (D'Souza 2007, pp. 263–264).

The British, however, frowned upon the so-called 'uncontrolled' systems of irrigation, such as flood inundation (Hardiman 2008, p. 50). Instead, water was to be managed according to the principle that nature could and should be mastered through science and technology. Natural resources should be, in Hardiman's words (2008, p. 50), 'transformed and made more productive, allowing Indian agrarian products to be sold competitively in a rapidly expanding world market'. Consequently, water resources became merely a commodity. Hardiman (2008, p. 50) writes that '[r]ivers were to be engineered, controlled, tamed and made into a source of water for artificial "rivers" (canals)'.

In the colonial period, then, the main purpose of water governance was that of canal irrigation, and water disputes meant disputes over allocation of water for agricultural purposes only (Iyer 2007, p. 152). This way of thinking has, according to Iyer, continued beyond British rule and is even enshrined in India's Constitution, which identifies water as rivers only (while ignoring groundwater) and is preoccupied with irrigation. There is no evidence of recognition of water as an integral part of the ecological system (Iyer 2007, p. 152).

The supply-based paradigm of water management carried on into the post-colonial period. The independence of India in 1947 coincided with improvements in engineering and a worldwide enthusiasm for dam-building. Big dams and mega engineering projects were not only regarded as a symbol of modernity and development in South Asia, but were also the favoured means of meeting water needs of the region (Asthana & Shukla 2014, p. 61). The apparent ease of water supply that dams offered appealed especially to Indian farmers, who created political pressure for constructing dams for irrigation purposes (Attwood 2007, p. 30).

This was also the era of massive irrigation schemes and other agricultural reforms in the Soviet Union, which greatly influenced the newly independent India. As Dhaka illustrates, poverty and famine were addressed 'through prodigious five-year planning and infrastructure development model envisioned along socialist path of reconstruction'. In this regard India took Central Asia as a noble example of social and economic transformation at a single go (Dhaka 2005, pp. 202–203).

Surface and groundwater irrigation increased enormously in India in the mid-twentieth century, although, as Hill (2013, p. 246) points out, two-thirds of the population continues to depend upon rain-fed agriculture as a direct or indirect source of livelihood. Rain-fed agriculture is especially important in the mountainous north where it is difficult to build irrigation infrastructure, and also the parts of the country that continue to have the largest concentrations of chronic poverty. Nonetheless, the expansion of the canal systems for irrigation facilitated the Green Revolution in the 1960s and 1970s (Hill 2009, p. 98; Hill 2013, p. 246).

The Green Revolution was an unprecedented boost to food production made possible through the availability of new, high-yielding crop varieties that required intensive irrigation. This, as Asthana and Shukla (2014, p. 82) argue, had a positive effect not only on food security but also on the regional economy; agriculture was no longer largely for subsistence purposes, but was able to provide income through the sale of excess produce. The Green Revolution, combined with the dam-building that expanded irrigation networks, increased agricultural output and allowed India to decrease dependence on food imports from abroad (Wood 2007, p. 99).

In addition to these on the whole positive developments, the Green Revolution also had the unintended consequence of creating an over-reliance on pumped groundwater to irrigate fields (Hill 2009, p. 98). A quarter of India's crops are now grown using non-renewable underground water (Pearce 2006, p. 25), and the area irrigated with pumped groundwater exceeds the area irrigated by surface water (Mollinga 2006, p. 33 footnote 6). This predicament is facilitated by universal access to cheap water pumps, Pearce argues (2006, p. 36). The availability of water pumps, claims Hill (2009, p. 99), combined with lack of regulation on groundwater extractions and largely unsatisfactory provision of municipal water services have facilitated a market in water-selling by local pump owners, who themselves are often former farmers.

The unsustainable exploitation of groundwater resources is a grave issue for India, but as it is only tangential to transboundary water governance in the Ganges–Brahmaputra problemshed, it is not addressed in greater detail in this book. Rather, the salient point here is that India, and perhaps the broader region, is in need of what Pearce (2006, p. 308) calls a 'Blue Revolution' of water use efficiency and sustainability. This could be an opportunity for foreign-led capacity building initiatives, and indeed some of the explicit objectives of international organisations and aid donors in the Ganges–Brahmaputra problemshed are to improve sustainable development of the region's freshwater resources. This is discussed in greater detail in Chapter 6.

Narratives of water scarcity and supply-side solutions

Mahatma Gandhi famously said that nature provides enough for everyone's need, but not for everyone's greed. This aphorism is often raised in relation to water governance in South Asia, usually to support arguments such as those made by Iyer (2007, pp. 41–42) that water resources are plentiful but being mismanaged or apportioned unjustly. It has also been invoked to support the discourse of water scarcity in the region – the discourse that claims there is simply not enough freshwater to satisfy the needs of all. This is the prevailing narrative in South Asia: water resources are seen as increasingly scarce, rather than increasingly strained and possibly mismanaged due to growing demand and competing water use priorities (Hanasz 2014a).

The narrative of scarcity has some consequences for water governance. Seeing the water crises in the region as primarily crises of *availability* implies that the solution, at least according to engineers and bureaucrats such as those at the World Bank, lies in increasing water 'production' through additional water infrastructure. In this paradigm, the quantity of water is the central problem rather than the clever or equitable distribution of it.

A supply-side approach to water resource management implies that solutions to water problems will focus on creating more water availability. This, however, can lead to a destructive, self-perpetuating cycle. As Iyer (2010, pp. 79–80) explains, there is always demand for more water but where will this 'more water' come from? It has to be brought from somewhere, so big dams, canals and long-distant water transfers are built. These, in turn, generate new conflicts. For Iyer, what lies at the heart of water conflicts is 'greed' – constant demand for more and more. Agreements, accords, treaties and adjudications may, he argues, temporarily bring peace, but disputes will continue unless the idea of 'development' is reconceptualised. Part of that reconceptualisation would be a move away from supply-side approaches to water governance.

According to Iyer (2007, p. 18), another solution favoured by the World Bank is in 'water markets' which allow for free trading that creates a supply-side response to the growing demand. As Hill (2015, p. 729) argues, this neo-liberal approach ignores the significant social, economic and political issues associated with this supply-side, scarcity-focused approach to development of water resources in the region.

One way in which this dominant water governance paradigm could be flipped is by reframing a problem of 'insufficient water supply' as one of 'excessive water consumption', thereby leading to solutions that favour the way in which water is allocated, used and consumed, rather than technological solutions that 'create' more water (Patrick et al. 2014, p. 11). Shifting the water governance paradigm in the Ganges–Brahmaputra problemshed from a supply-side approach to a demand side is one of the approaches identified by riparian stakeholders (and discussed in Chapter 12) as desirable for improving transboundary water governance.

As it currently stands, however, water governance in South Asia is dominated by supply-side concerns, and by the various disciplines of engineering, claims Routray (2010, p. 34). The National River Linking Project is one prominent example of this within India (Iyer 2007, pp. 46–57). In terms of transboundary water governance in the Ganges–Brahmaputra problemshed, it too has, according to Prasai and Surie (2013, p. 1), 'long been dominated by technical perspectives from civil engineering, economics, and international law'. Hill (2015, p. 741), too, sees the dominance of these perspectives in the region, especially in regard to hydropower development. Hydropower development, he argues, is overwhelmingly guided and constrained by the supply-side paradigm.

The narratives of scarcity and supply-side solutions also frame water as a finite and single-use resource. This framing creates a zero-sum attitude to water sharing. That is, in the narrative of scarcity, the idea that water can be used for multiple purposes, and create multiple values from the same quantity, is lost. As Paranjpye (2010, p. 124) shows, even at the domestic level in India there is a 'lack of integrative thinking among all parties concerned'. This is also a feature of transboundary water interactions in the Ganges–Brahmaputra problemshed. As Wirsing et al. (2013, p. 13) point out, '[a] zero-sum water resource atmosphere is clearly building up in Himalayan Asia'.

Multi-scale, non-water related priorities

Another important contextual factor is that water interactions exist in a complex political dynamic that operates on various but interconnected scales (local, national, regional, global). These multi-scale socio-political dynamics may be a distraction from, or prioritised higher than, issues of transboundary water governance. As Hill (2008, p. 75) makes clear, changes in the domestic political scene in India, Nepal, Bhutan and Bangladesh impacts transboundary water policy priorities.

In India, for example, inter-state water conflicts are more pressing than transboundary ones (Hanasz 2015b). Bangladesh is focusing on rapid economic growth rather than sustainable development of natural resources. Meanwhile, Nepal has been grappling with an insurgency, political instability and a lack of domestic consensus on how to develop that country's vast hydropower resources (Hanasz 2015a). Only Bhutan seems to have sufficient political will to satisfactorily address its transboundary water issues with India, but that may be because it is a small, politically stable, ethnically and religiously homogenous country with few options for economic development other than by selling hydropower to India.

Hill (2015, p. 737) argues that India's bureaucracy is also a limitation on transboundary cooperation. He claims that the bureaucratic culture prevailing in India is a bigger obstacle to effective water resource management than disagreements between states. He quotes the World Bank, which suggests this bureaucratic culture is characterised by 'the paternalism of central-level bureaucrats, coercive top-down planning, and little support or feedback from locals'.

Hill's reference to a recent World Bank report illustrates that the World Bank itself is not blind to all contextual factors affecting transboundary water governance in South Asia. As I argue, however, this understanding of the region does not translate into tailored solutions to water governance problems; foreign expertise and one-size-fits-all approaches are favoured by international actors above context-specific ones.

The disagreements between states in India are a further limitation on the development of the region's water resources. There are disputes about all the major rivers in India but, according to Paranjpye (2010, p. 121), the contending states 'do not appear to be seriously interested in finding solutions that are acceptable and beneficial to all concerned'. Instead, water disputes within India are, in Iyer's words (2007, p. 25), 'characterised by bitterness, tend to get enmeshed in party politics, and become intractable'. Overall, the domestic politicking over inter-state water disputes within India distract political attention from transboundary matters. As Dash (2008, p. 40) notes, '[r]egional cooperation is a two-level process in which domestic support and regional bargains and negotiations must overlap if cooperation is to proceed'.

Domestic issues, however, simply take precedence over regional cooperation in the Ganges–Brahmaputra problemshed. According to Dinar (2002, pp. 246–247) domestic politics play a large role in the hydropolitics between the basin riparians. Cooperation between Bangladesh, Nepal and India is curtailed by domestic political factions that accuse their ruling parties of compromising their nation's sovereignty and national interest. He argues that '[a]lthough hegemonic stability theory tells us that a lack of regional cooperation is due to the desires of India to prevent multilateral regional cooperation, domestic factors including political instability and nationalist fervor intensify the conflict and also prevent collaboration'.

A further effect of the intense politicking between Indian states over water is that it creates an unhelpful impression. Prasai and Surie (2014) point out that there is a perception that if India cannot manage its own domestic water conflicts, it will not be able to manage the international ones. Also, there is an assumption that 'conflicting needs and interests of the different states must be reconciled domestically before any international agreement can be reached' (Dinar et al. 2007, p. 255). Yet, India's transboundary water relationships are better than the water interactions between Indian states (Bhaduri & Kaushal 2014). While a degree of basin-wide coordination and integration exists between India and its neighbours, Iyer (2007, pp. 113–114) notes that no such arrangements exist *within* India and that there is strong resistance to the idea of the River Basin Organisation (RBO) on the part of the state governments.

Downstream of India on both the Ganges and Brahmaputra Rivers is Bangladesh. This country is especially vulnerable to any quantitative and qualitative impacts caused by the actions of upstream riparians because approximately

94 per cent of its surface waters originate outside Bangladesh (Dinar et al. 2007, p. 252). As such, it seems reasonable that Bangladesh should devote a significant amount of political energy to maintain cooperative relations with India, and ensure that any upstream actions do not create adverse consequences within Bangladesh. This, however, is not the case. Certainly, a high priority for Bangladesh is developing a constructive water policy and attaining development goals associated with this (Condon et al. 2009, p. 41). But a national water policy establishes how water is to be managed domestically, not how important transboundary waters are.

Yet, transboundary waters are not *unimportant* to Bangladesh. Begum Khaleda Zia, the former Prime Minister of Bangladesh, has said that 'the sharing of the waters of our common rivers' is one of the most pressing issues vis-à-vis India (Zia 2012, p. 721). But while water disputes are high up on the bilateral agenda for Bangladesh, they are not a relatively high priority overall in the context of other social, political and economic issues on various levels. The political situation in Bangladesh remains unstable (World Bank 2014b, p. 3), while rapid economic growth currently trumps environmental considerations and sustainable development. These factors stand in the way of Bangladesh resolving disputes with India, for example over the Teesta River.

Nepal too is aggrieved by India in regard to transboundary water governance in the Ganges–Brahmaputra problemshed, and like Bangladesh is not able to focus political will on resolving these grievances – but for different reasons. Nepal continues to feel cheated over existing hydropower project deals with India, even ones dating as far back as the 1950s, and the bitterness generated by these experiences has coloured all subsequent water interactions between the two countries (Bhattarai 2005, p. 15). The persistent resentment towards India hinders the cooperative development of shared water resources.

Decades of political and social instability in Nepal have also detracted from the political attention available to address transboundary water management issues (Chellaney 2011, p. 283). Nepal has been wracked by severe political flux since the early 1990s (Chellaney 2011, p. 283) and it is now recovering from decades of insecurity and instability. The state's failure to initiate socio-economic development is one of the drivers of conflict in Nepal, while the chaos and violence in turn hinders the state's ability to initiate socio-economic development (Bhattarai 2005, p. 28). Most development activities halted in rural Nepal, and as a result of civil war the government slashed development funds, diverting them to security expenditures. Moreover, rebel attacks on hydropower plants, access facilities, communication networks and development projects caused substantial economic losses and general devastation (Baral & Heinen 2005, p. 3).

At the same time, as Gyawali and Dixit (1999, pp. 559–560) argue, Nepali party functionaries fear political allegations that a project, any project, is not moving forward due to their inaction regardless of technical, economic or developmental demerits of such projects, and so they prefer to be seen as 'for' projects rather than questioning them. There is enormous political pressure within Nepal to harness water resources vis-à-vis India, particularly through mega projects, which are believed to bring more economic benefit and political kudos than more moderately sized undertakings (Gyawali & Dixit 1999, p. 563). But the real problem lies, according to Shakya (2009, p. 72), in the lack of political consensus on any projects, big or small.

In contrast to the numerous issues that distract and divide political will in Nepal away from transboundary water governance is the case of Bhutan, which has made a concerted effort to come to terms with India on issues of mutual interest. The fact that Bhutan is small, ethnically, religiously and politically homogenous, with a stable government, and few other socio-economic issues to distract it, means that it has been able to channel political will into resolving contentious water resource development issues with India. This arrangement is positive-sum and the export revenue that Bhutan receives from the sale of hydropower to India is not only a major source of GDP but has also fuelled remarkable socio-economic growth (Hanasz 2015a).

While the quality of water interaction between India and Bhutan provides an interesting juxtaposition to the hydropolitics between India and Nepal, there is no meaningful comparison between the two countries: Bhutan is one fiftieth the size of Nepal, has no Terai lowlands to irrigate (Gyawali 2013, pp. 7–9), and is governed by a stable central government. The 'Bhutanese model', if there is one, is therefore not appropriate for the Nepal case. The core water issue between Bhutan and India is hydropower as Bhutan is upstream of the water-rich state of Assam (fed by the Brahmaputra River), while Nepal is upstream of the water-scarce and populous states of Bihar and Uttar Pradesh, thus making irrigation and flood control significant issues to resolve with India (Gyawali 2013, pp. 7–9). The purpose of juxtaposing Nepal and Bhutan, then, is to illustrate that water interactions are complex and determined by different factors on different scales, and not just riparian position vis-à-vis the hydro-hegemon.

In addition to these domestic issues within Bhutan, Bangladesh, Nepal and India, there are also international dynamics that affect the hydropolitics between them. As Earle et al. (2015, pp. 41–42) point out, China's influence on the hydropolitics of the Ganges–Brahmaputra–Meghna basin brings further complications and uncertainties. Although it is not within the scope of this book to consider China in greater detail, the point remains that how India, Nepal, Bhutan and Bangladesh relate to and are influenced by China affects the hydropolitics between them.

These examples of domestic barriers to developing transboundary water cooperation illustrate that dispute resolution mechanisms and institutions alone are not immune to politics at various scales and that the political context must be considered when addressing water conflicts. This point recurs in Chapter 10, which examines foreign-led interventions in building transboundary water cooperation in the Ganges–Brahmaputra problemshed: the lack of consideration of the political context is a significant weakness of these interventions. The lack of priority given to transboundary water issues by the riparian states in the problemshed is a further limitation on the likelihood of transboundary water interaction in the Ganges–Brahmaputra problemshed improving significantly in the foreseeable future.

Challenges for transboundary water governance

The riparians of the Ganges–Brahmaputra problemshed may be preoccupied with issues other than transboundary water cooperation, but that does not mean that there are no crises or challenges for shared water resources that require policy attention in this region. Some of these challenges are similar the world over: population growth, rise in demand for food, energy and water, as well as rapid urbanisation. This is in addition to the effects of climate change, which may be particularly acute in this poor and densely populated region that is also not especially resilient in terms of its institutions and polity.

These factors, which are not unique to the Ganges–Brahmaputra problemshed, have given rise to the fear of socio-political conflicts, including international disputes, over increasingly scarce natural resource and human displacement due to environmental degradation. These fears, in turn, drive international organisations and bilateral aid donors, among others, to address the many water crises the world over to mitigate the risk of such conflicts developing.

These global pressures drive increasing demand for freshwater. That, in turn, will require ever greater diplomacy to, as Verady et al. (2014, p. 28) argue, anticipate, mitigate and resolve conflicting claims on shared water resources. This will be particularly important in transboundary basins, because states will need to balance national interests with regional welfare and economic security: 'Therein lies the great challenge of hydro-diplomacy and the concomitant potential of hydro-solidarity' (Verady et al. 2014, p. 28). Climate change, population growth and urbanisation will be, according to Keskinen et al. (2014, p. 37), the 'hard' drivers of transboundary water cooperation, 'as opposed to the sometimes perceived notion of cooperating on "soft" environmental values'.

This section outlines how these 'hard' drivers play out in the Ganges–Brahmaputra problemshed and the challenges they pose to transboundary water governance. First are the connected challenges of economic development, rapid population growth and urbanisation. Next are the effects of climate change, followed by the issue of food security.

Economic development, rapid population growth and urbanisation

The greater Himalayan region supports nearly half of the world's population, and it is very densely populated. The Ganges–Brahmaputra–Meghna basin is home to 630 million people in an area of 1.7 million square kilometres (Vaidya & Sharma 2014, p. 17). India today has to support over 17 per cent of the world's population on just over 2 per cent of the world's land area, while Bangladesh supports about 2.27 per cent of the world population on 0.11 per cent of the world's land area, and these ratios are bound to widen by mid-century (Wirsing et al. 2013, p. 14). India is expected to overtake China as the world's most populous country by the year 2035 (Kumar & Furlong 2012, p. 5). The implication of this population growth for water management is grave: more people means more demand for food, energy and water.

Asia is already the world's most water-stressed continent. The per capita availability is less than 1,700 cubic metres per year, the internationally accepted definition of water stress (Chellaney 2014, p. 622). In India, the richest country in the Ganges–Brahmaputra problemshed, per capita availability of water is expected to drop below 1,000 cubic metres (that is, below officially accepted water scarcity levels) by the end of this century if the pace of development continues at current levels (Kumar & Furlong 2012, p.5). Asia's per capita water availability is the lowest of any continent, and its water stress has been exacerbated by a dramatic economic growth coupled with rapid urbanisation (Chellaney 2014, p. 622).

Increasing populations, urbanisation and development have also been identified in India's National Water Policy (in 2012, but also in previous years) as a major problem for water governance. Rapid urbanisation alone 'is driving increased water demand both for municipal supply and for the industrial and agricultural products in demand in cities' (Chellaney 2014, p. 622). Even in sparsely populated Bhutan, which is home to only 733,000 people, rates of urbanisation are higher than in India, Nepal or Bangladesh (CIA World Factbook 2015), and the Royal Government of Bhutan admits that '[u]rbanisation has become a key issue that has serious impact on both water demand and quality' (Royal Government of Bhutan 2003, p. 3).

Iyer (2008b, p. 7) cautions, however, that while large supply-side projects and long-distance water transfers may seem like obvious solutions to these problems, they are in fact only band-aid solutions that may exacerbate conflict in the long term. As discussed above, supply-side, engineering and economically driven solutions already dominate the transboundary water governance regime in the Ganges–Brahmaputra problemshed, and so growing pressures may further cement these approaches rather than encourage demand-side or innovative policy solutions.

Climate change

Water management issues stemming from population growth, rapid urbanisation and socio-economic development are further exacerbated by the effects of climate change. The countries of the Ganges–Brahmaputra problemshed are less resilient to water-related disasters than other Asia-Pacific states, which creates a further challenge for transboundary water governance and climate change adaptation (Vaidya & Sharma 2014, p. 12). As Hill (2015, p. 739) notes, the phrase 'too much and too little water' has been used as a shorthand for the challenges that climate presents for the Hindu Kush Himalaya. Droughts and floods are increasingly more commonplace. The region is also ill-prepared institutionally for climate change. For example, none of the treaties and agreements that Nepal has with India address climate change or the uncertainty posed by potential effects of changing melt dynamics from Himalayan glaciers (National Research Council of the National Academies 2012, p. 86).

The effects of climate change include complications with spatial and temporal rainfall variability, evaporation rates, and temperatures in different agro-climatic zones and river basins (Asthana & Shukla 2014, p. 139). This, in turn, will affect 'agricultural production and food security, ecology, biodiversity, river flows, floods and droughts, water security, and human and animal health' (Vaidya & Sharma 2014, p. 17). Changes in climatic conditions also affect diseases transmitted through water, via vectors such as mosquitos (Werz & Conley 2012, p. 20).

The compounded effects of rising temperature, flooding, sea level rise, loss of wetlands and cyclones (Evans 2015b, p. 80) will leave densely populated 'mega-deltas', such as the Ganges–Brahmaputra–Meghna basin particularly at risk (Evans 2010, p. 4). In this region, climate change will likely result in increasingly frequent and hazardous weather patterns, such as extreme temperatures, changes in average annual precipitation, irregular rainfall patterns and intense rainfall events. These, according to research from the International Centre for Integrated Mountain Development (ICIMOD), may in turn lead to increases in floods, landslides, and erosion, as well as an increase in the frequency and intensity of droughts. The result will likely be lowering of the water table, drying of natural springs, and reduced stream discharge (Vaidya & Sharma 2014, p. 20).

The large number of dams in the region also poses a unique problem in light of climate change: With the accelerated glacial melting, Kumar and Furlong (2012, p. 10) argue that the dams are likely to see huge increases in inflows initially and then highly reduced inflows in subsequent decades. The safety and the economy of the dams is thus threatened. At the same time, climate change and the priority placed internationally on reducing greenhouse gas emissions has led, since the World Commission on Dams report, to rekindled interest in hydropower (Smith 2010, p. 440). That means more dams – and more dams means more risk associated with climate change.

Water supply will also be affected by climate change. Evans (2015b, p. 80) points to out that water for irrigation is threatened, as is the water-energy nexus: water for hydropower production and the cooling of thermal power production are also at risk. Moreover, the annual mean monsoon levels will increase by 10 percent, with a 15 percent increase in variability. This, argues Evans, will make the monsoon stronger and less predictable. Climate change will likely affect seasonal water supplies more than annual water supplies (Vaidya & Sharma 2014, p. 15). In the Ganges basin, dry season, monsoon and post-monsoon flows are projected to increase by up to 20 per cent in the 2050s, while pre-monsoon flow is expected to decrease by 15 per cent; in the Brahmaputra and Meghna basins, flow is projected to increase in all seasons, with a marked increase in the Brahmaputra basin in the dry season (17 per cent), and in the Meghna basin in pre-monsoon (20 per cent) (Vaidya & Sharma 2014, p. 17). Studies also indicate that mean upstream water supply on the Ganges and Brahmaputra in the period 2046–2065 will be approximately only 80 per cent of what it had been in 2000–2007 (Vaidya & Sharma 2014, p. 15).

Much of the dry season flow of Himalayan rivers is from glacial meltwater, but over the last 30 years this has fallen by 11 per cent as average temperatures rise and snowlines retreat (Evans 2010, p. 12). In contrast to Europe and North America, however, where glacial melt contributes substantially to summer flows, the glaciers of the greater Himalaya region melt during the monsoon season when temperatures are highest but rainfall is also heaviest. Thus, while changes in glacial melt may be an existential challenge for some melt-dependent mountain communities, it is not a major challenge for the problemshed as a whole (World Bank 2014a, p. xvii).

Nonetheless, a major concern is that the rapid retreat of glaciers in recent decades has resulted in the formation and growth of glacial lakes. These are held back by unstable moraine material and are prone to burst in a type of flash flood called a Glacial Lake Outburst Flood (GLOF). Research from the International Centre for Integrated Mountain Development (ICIMOD) notes that there is likely to be more and more of this type of flood event due to climate change (Vaidya & Sharma 2014, p. 20). Regular floods are also expected to increase in frequency and severity, and to alternate with drought (Vaidya & Sharma 2014, p. 9). Protracted droughts combined with rising sea levels may cause inland backflow of salt water that contaminates low-lying, fertile delta regions (Campbell et al. 2007, p. 75).

Despite these gloomy predictions, some international actors remain optimistic that the effects of climate change in the Ganges–Brahmaputra problemshed will be manageable, and – crucially for proponents of dam-building in the Himalayas as a means of controlling water – that climate change will not adversely affect the hydropower potential of the region. Jeuland et al. (2013, p. 26), who were all involved with the World Bank's Ganges Strategic Basin Assessment, argue that predicted future flows are not outside the range of natural variability in the basin. More importantly, they find that climate change will likely not adversely affect the hydropower potential of 23 large dams in Nepal because annual flow in the tributary rivers greatly exceeds the storage capacities of these projects even in dry scenarios (Jeuland et al. 2013, p. 26).

International actors, however, are not united in calling for infrastructure to regulate the vagaries of water flow in the Ganges–Brahmaputra problemshed. The Ganges Strategic Basin Assessment, which was made public more than a year after Jeuland et al. published their assessment of climate change on hydropower development in the region, concludes that 'the effectiveness of large-scale infrastructure for flood control, and the reliability of existing large-scale diversions of surface water for irrigation, could prove susceptible to climate change' (World Bank 2014a, p. xvii). The report argues that improved policy is key to rising to the challenge of climate change: 'Regardless of changes in rainfall and hydrology, an emphasis on enhanced forecasting and warning systems, in concert with a suite of tailored, localized responses, is urgently needed' (World Bank 2014a, p. xvii).

Yet, this policy-centric approach to hydrological variability fell on deaf ears as the Ganges Strategic Basin Assessment was rejected by Nepal, India and

Bangladesh. This suggests, among other things, that supply-side, engineering solutions are deeply entrenched in this problemshed.

Food security

A further predicted effect of climate change is on food security. Saltwater inundation and increased coastal erosion will affect food production (Werz & Conley 2012, p. 7), as will rising temperatures (Evans 2015b, p. 80). To maintain crop production levels that counterbalance these effects and are sufficient to feed growing populations, annual water withdrawals for agriculture are projected to increase in South Asia by 9 per cent from 2000 to 2050 (Vaidya & Sharma 2014, pp. 6–7). This will further exacerbate the pressures on and competition for the finite and stressed freshwater resources of the Ganges–Brahmaputra problemshed. Furthermore, the nexus of food and water security is particularly worrisome because pressure in just one factor, such as decreased crop yield, can continue to have cascading effects on populations for generations, such as malnutrition and stunting in childhood (Evans 2015b, p. 80). Evans (2015b, p. 80) predicts that decreasing food availability in 2050 will cause a 35 per cent increase in childhood stunting as undernourishment worsens.

Agriculture absorbs 90 per cent of the total usage of water and remains the most significant factor in food security and rural livelihoods throughout India (Hill 2013, p. 346). Yet it was not until 2012 that India included food security within its National Water Policy (Ramesh 2012). And, as Prasai and Surie (2013, p. 1) observe, the lack of regional cooperation in the Ganges–Brahmaputra problemshed continues to prevent sustainable development and management of transboundary water resources for food security and livelihood improvement. Nonetheless, international organisations and foreign aid donors understand that '[t]o ensure sustained food security, economic growth and political stability in the future, water must be managed better' (Rollason 2012). This, and other objectives of international actors, are examined in Chapter 6.

3 The transboundary water governance agenda in the Ganges–Brahmaputra problemshed

In the context of water-related legacies and challenges discussed in Chapter 2, there are a number of specific issues for transboundary water governance in the Ganges–Brahmaputra problemshed. Each presents unique risks and opportunities, but they are largely interrelated and, as with any wicked problem, addressing one issue in the problem set may adversely affect another. Considering this complexity, it may be difficult to prioritise which issues are most pressing. Yet, the lack of policy priority and political will on any one issue between India, Nepal, Bhutan and Bangladesh also means that none of the issues receive adequate attention. Without a single crisis to galvanise action, the many water crises remain governed on a 'business as usual' principle.

Part of the problem lies in the lack of systematic knowledge about the Ganges–Brahmaputra and Meghna basins. As the World Bank (2014a, p. xiii) points out, information and data are scarce and difficult to obtain, and there has been no basin-wide knowledge base and analytical framework that could be used by riparian states to explore options and facilitate cooperative planning. To that end, the World Bank has taken on the task of documenting the water governance issues facing the Ganges and Brahmaputra Rivers. The 2014 Ganges Strategic Basin Assessment (GSBA) is a 'state of the basin' stocktake of hydrological, economic, social and political information, much of it based on primary research and the latest modelling techniques.

The study intended to dispel many commonly held assumptions and demystify the misconceptions that have driven freshwater development in the basin to date. It was to be the blueprint for a new era of evidence-led policymaking. In the absence of any other clear and consistent agenda for transboundary water governance between India, Nepal and Bangladesh, it was an attempt to focus effort and foment action on water issues of mutual concern, although this has not (yet?) occurred. The problematic issues with the Ganges Strategic Basin Assessment are discussed at length in Chapter 7. For the purpose of this chapter, however, the significance of the report lies in its systematic assessment of transboundary water issues.

The Ganges Strategic Basin Assessment is also a *de facto* catalogue of what the World Bank considers the most pressing issues for this basin – water storage, water for irrigation, hydropower generation, flood mitigation and sedimentation. These issues are also of concern in the Brahmaputra basin (although at time of writing the Strategic Basin Assessment of Brahmaputra System in Northeast India, the companion report to the Ganges Strategic Basin Assessment, was not complete). Each of these issues is discussed in turn, first from the perspective of the riparian states, and second from the perspective of the World Bank.

Upstream water storage: the view from riparian states

Considering the importance of agriculture in the Ganges–Brahmaputra problemshed, it would be logical to assume that the most pressing issue for water governance is that of irrigation. If not irrigation, then flood control and mitigation could be the main concern considering the effects of climate change and the large, dense and poor population of this region. Hydropower production could also be considered a high priority because of the rising energy demand. Underlying these three issues, however, is the problem of water storage.

The Hindu Kush Himalaya mountain range, from which the Ganges and Brahmaputra Rivers flow, holds one of the largest bodies of ice outside the polar caps, and these glaciers, ice fields and snow packs provide important intra- and inter-annual water storage facilities. During the summertime monsoon, however, the contributions of meltwater and rainfall coincide, creating a situation of too much water in the wet season and too little water in the dry season. The increasing gap between water availability and demand in months other than the monsoon season is threatening livelihoods and economic development in the region, but could be alleviated with man-made water storage facilities (Vaidya & Sharm 2014, p. 3).

As analysts from the International Centre for Integrated Mountain Development (ICIMOD) argue, water storage is increasingly important in the Hindu Kush Himalayan region. It can be used to even out supply in areas where the intra-annual precipitation is uneven. It can also be used, they claim, as a buffer against high precipitation events, and it can be used to store water at times of low requirement for use at times of high requirement. As such, water storage may also become a key strategy for climate change adaptation (Vaidya & Sharm 2014, p. 11).

As the Ganges Strategic Basin Assessment suggests, the storage of water high in the Himalayas could have profound benefits far downstream: 'The effects of increased low flows may make important contributions in the Ganges delta areas to better manage saline intrusion, enhance the Sundarbans ecosystem, and maintain navigation services' (World Bank 2014a, p. xv).

Water storage is a particularly important transboundary issue because it can be used to even out water supply between where it is comparatively plentiful – Nepal and Bhutan – and where it is scarce – India. India's per capita yearly water storage (200 cubic metres) is low relative to global averages (Chellaney 2011, p. 89), and approximately 35 per cent of water needed for irrigation comes from Indian reservoirs (Bandyopadhyay et al. 2002, p. 4108). The remainder of India's irrigation needs are fulfilled by groundwater, thus contributing to the overexploitation of aquifers. The storage of monsoon runoff in the Nepali hills could, according to Gyawali (2013, p. 6), replace groundwater for irrigation in the northern Indian states of Uttar Pradesh or Bihar.

In addition to flow augmentation for downstream irrigation, Nepal's storage potential of 88 bcm could bring other mutual benefits, such as hydropower generation, flood control during the monsoon, flow augmentation for navigation and thus, in Onta's words (2001, p. 106), 'it is natural that India has a vital concern about Nepalese waters and will have keen interest in Nepal's water resources development'.

At the same time, Nepal's lack of diversion or storage capacity is at the root of water disputes with India, according to Condon et al. (2009, p. 13). More specifically, the crux of water disputes between India and Nepal lies in how each side values the benefit sharing options arising out of potential water storage. As Gyawali (2013, p. 6) argues, even though Nepal's primary interest is in hydroelectricity, it is also very much in the national interest to make sure that the cost of a dam is paid for by the other sector beneficiaries, that is, flood control, environmental flows, navigation and especially irrigation.

While for Nepal the issue of water storage is largely one of untapped potential, for Bangladesh it is a question of mitigating the negative effects of India's upstream projects. Bangladesh has a seeming abundance of water resources, but most of its annual flow comes from monsoonal floods that quickly flush out to the Bay of Bengal because Bangladesh is a predominantly flat, deltaic country with limited water storage potential (Wirsing et al. 2013, p. 15). At the same time, Bangladesh has not benefitted from water storage and diversion in neighbouring India.

The Farakka Barrage on the Ganges River in the Indian state of West Bengal is a particular and ongoing point of contention between the two countries. Indeed, as Hill illustrates, the tension over the Farakka Barrage is symptomatic of the broader relationship between India and Bangladesh. While India has aggressively asserted its own interest to the detriment of Bangladesh, claims Hill (2009, p. 92), on the other side of the border almost every negative development is linked to the Farakka Barrage in upstream India.

Both India and Bangladesh, however, agree that the 1996 Ganges Water Treaty, which regulates the division of water resources at the Farakka Barrage, does not guarantee sufficient water for either party and thus negotiations have shifted to the question of how to augment storage (Hill 2008, p. 73). Bangladesh proposed building storage facilities in the upper Ganges basin, that is, in Nepal, to collect monsoonal rainwater for release during the dry season (Tiwary 2006, p. 1688). This plan would also allow Nepal to generate hydroelectricity for sale to India, and provide some flood control benefits and enhanced inland navigation on the Kosi River, a tributary of the Ganges (Onta 2001, p. 110). India, however, avoids multilateral solutions and prefers to maintain strictly bilateral relations (this being the hydro-hegemon's prerogative, a situation examined in Chapter 9). In response to Bangladesh's plan, India proposed the construction of a canal to divert water from the Brahmaputra to the Ganges. As Tiwary (2006, p. 1688) explains, '[t]his suggestion was unacceptable to Bangladesh. Excavations would have caused loss of fertile land and the displacement of communities.'

The question of water storage is also on the domestic water governance agenda in India. As Kumar and Furlong (2012, pp. 12–13) point out, India's 2012 National Water Policy is based on the assumption that increasing water storage is the most favourable solution to various water problems. As discussed in Chapter 2, the water crisis in India is portrayed as a crisis of availability. The answer to this crisis, according to Iyer (2010, pp. 79–80), is to increase water availability through supply solutions. In other words, increasing storage, mainly through large dams, and redistributing water through inter-basin transfers and long distance transfers from areas of 'surplus' to areas of deficit. The Policy itself makes this line of thinking explicit by determining that '[a]ll water resources projects, including hydro power projects, should be planned to the extent feasible as multi-purpose projects with provision of storage' (Government of India 2012, paragraph 9.7).

There are, however, critics of this approach to water storage in India. The South Asia Network for Dams, Rivers, and People (SANDRP), for example, calls this statement about multi-purpose projects 'dangerous'. Giving a privileged position to mega storage options is uncalled for, according to SANDRP, because other storage options (e.g. increasing soil capacity to retain moisture, local storage, underground storage) and optimum use of existing storage capacities can be explored (SANDRP 2012).

Furthermore, the importance put on supply side solutions and large infrastructure projects, according to Kumar and Furlong (2012, pp. 12–13), means that India has a blind spot to reducing pollution and demand. India apparently is neglecting small-scale conservation efforts such as water harvesting, and has done little to promote less water intensive crops, industry and energy plants. These points about the detrimental effects of a supply-side approach to water management, however, have not sparked a public discussion in the way that dams have. Hill (2013, p. 244) believes this is because storage and distribution of water are far more complex issues than that of hydropower.

Upstream water storage: the view from the World Bank

In the absence of a constructive dialogue between riparian states about water storage in the Ganges–Brahmaputra problemshed, the World Bank has begun to tease out the relevant issues and to separate myth from reality. The Ganges Strategic Basin Assessment identifies water storage as one of the most important issues for water governance in the Ganges basin, and one that could be the basis for creating positive-sum outcomes over other pressing issues – hydropower development, flood control, irrigation. Of the Assessment's ten key questions, five refer directly to water storage:

- Is there substantial potential for upstream reservoir storage in the Himalayan headwaters of the basin?
- Can upstream water storage control basin wide flooding?

- Can upstream water storage augment low flows downstream?
- Are there good alternatives or complements to reservoir storage?
- What are the cost- and benefit-sharing dynamics of upstream water storage development?

(World Bank 2014a, p. 49)

At the same time, the study interrogates the assumption that there is potential for large water storage in the Himalayan rivers and that 'this potential could be harnessed through large multipurpose dams to produce hydropower, deliver more timely irrigation water, and regulate the extreme flows of the Ganges River' (World Bank 2014a, p. xiv). The findings of the Ganges Strategic Basin Assessment refute the assumption that upstream reservoirs in Nepal can control basin-wide flooding. However, the study did find that such dams could potentially double low flows in the dry months. But the value of doing so is unclear and, the World Bank argues, similar storage volumes could be attained through better groundwater management (World Bank 2014a, p. xiv).

The analysis of the World Bank team suggests that very little water can in fact be stored in the steep terrain and mountain gorges of the region, even behind high dams. Developing all the structures currently proposed would provide only about 18 per cent of the basin's annual average flow, which is very little storage on a basin-wide scale (World Bank 2014a, p. xiv).

Flood control is also shown to be a negligible benefit of increased water storage in the Himalayas. Large dams in the region are, according to the Ganges Strategic Basin Assessment, commonly seen as the answer to the flooding that plagues the Ganges plains and delta. Yet, the potential to control floods throughout the basin using upstream storage is limited. This 'severely constrains' riparians' ability to ever truly regulate this river system, even assuming an aggressive development of system storage (World Bank 2014a, pp. xiv–xv). On the other hand, the lack of substantial regulation and water storage will, according to the Ganges Strategic Basin Assessment, preserve a more natural hydrology in the river system. Natural hydrology provides a wide variety of services that have not been quantified in this report, such as ecosystem services and navigation (World Bank 2014a, pp. xiv–xv).

A further assumption about water storage is that it could be used to retain wet season rainfall upstream for use in the dry season downstream. The Ganges Strategic Basin Assessment largely confirms this logic, albeit with some caveats. The construction of all currently planned dams could almost double flows during the months with the lowest flows, but while this may be large relative to current low flow, it is negligible compared to peak flow. Thus dry season flow augmentation through increased water storage is unlikely to significantly alter the integrity of the Ganges basin hydrological system as it currently stands (World Bank 2014a, p. xv). Moreover, 'the economic value of this low-flow augmentation is unclear because of low agricultural productivity and localized waterlogging' (World Bank 2014a, p. xv). In other words, agricultural modernisation will be required to increase productivity, and this modernisation will be beneficial irrespective of upstream dam construction (World Bank 2014a, p. xv).

The point that other strategies will have to complement or maximise the benefits of water storage in the Himalayas is strongly made in the Ganges Strategic Basin Assessment. The study puts forth the argument that there are many attractive alternatives to meeting the region's water needs; constructing new water storage infrastructure is not the only option. This is a novel and important argument in a water governance context dominated, as discussed in Chapter 2, by the preference in the region for engineering solutions to water supply policy problems. It is not one, however, that has (yet?) been embraced by riparian states.

The Ganges Strategic Basin Assessment shows that contrary to the prevailing command-and-control paradigm, there are numerous forms of water storage already present in the Ganges basin that are not human-constructed, namely underground aquifers, lakes, glaciers, snow, ice and even soils. The quantity of water held in these natural reservoirs may not be as insignificant as is widely believed (World Bank 2014a, pp. xv–xvi). The study reveals that in contrast to the over-abstraction of groundwater elsewhere in South Asia, there are vast, untapped groundwater resources in the central and lower reaches of the Ganges Basin, and that these can be sustainably used (World Bank 2014a, pp. xv–xvi).

More importantly, the sustainable use of this groundwater in strategic conjunction with well-managed surface water, could provide the same benefits, and on a similar scale, as those that could be gained from the current and proposed multi-purpose dams. But that, argues the World Bank, requires policy reform, and changes in farmers' behaviour (World Bank 2014a, pp. xv–xvi). Yet, policy reform, in combination with cultural and attitudinal change, is more difficult to bring about and measure than storing water behind multi-purpose dams.

With almost no flood control benefits, and the benefits of dry season flow augmentation being limited by low agricultural productivity, the overwhelming share of economic benefits from proposed dams will be, according to the Ganges Strategic Basin Assessment, derived from hydropower. The benefit-sharing calculus is thus simple: the benefits and costs to be shared between Nepal, India and Bangladesh in the near term will be predominantly associated with hydropower (World Bank 2014a, p. xvii). As Kayastha (2001, p. 142) argues, these countries are less concerned with the other benefits of dams – irrigation, flood control, navigation – than they are with hydropower production, and thus agreement must be reached on cost-sharing for this benefit before the other benefits can be realised.

Hydropower generation: the view from riparian states

Gyawali takes a different view to Kayastha in regard to the cost-sharing and benefit-sharing of Nepal's hydropower potential. He argues that even though Nepal's primary interest is in hydropower development, *all* potential costs and benefits must be accurately calculated so that there are no 'free riders' (Gyawali 2013, p. 6). The lack of consensus on how to share the benefits and the costs of developing transboundary water resources lies, according to Gyawali, at the heart of disputes over dam-building. There is no one right way of benefit sharing.

Tensions between riparians in the Ganges–Brahmaputra problemshed continue to fester over existing and planned hydropower projects. Hill (2013, pp. 250–251) argues that future large-scale hydropower development will increase tensions across the Hindu Kush Himalaya, perhaps to a greater extent than projects such as the Farakka Barrage (West Bengal), or Mahakali Irrigation Project (Nepal) did in the past. This accords with Chellaney's (2011, p. 25) view that '[t]he construction of dams on international rivers to generate hydropower is often the principal cause of water-related tensions between neighbours in Asia'.

Nonetheless, the development of Himalayan hydropower potential is the primary pursuit of hydrocrats and investors in the Ganges–Brahmaputra problemshed, especially in Nepal. India is the main beneficiary of these investments. It is said that water and energy pose the biggest constraint on India's growth because demand for both is increasing at a rate faster than current capacity can provide (Kumar & Furlong 2012, p. 8). Part of India's energy shortfall can be addressed through domestic hydropower generation (Chellaney 2011, p. 178), though there is some civil society opposition to dam building in India (Bharadwaj et al. 2006, p. 1206). Nepal's enormous hydropower potential could provide a convenient and significant supply of 'clean' energy for India's growing needs, particularly in the north of the country (Onta 2001, p. 110).

Nepal's theoretical hydropower potential is enormous, at 83,000 MW (although the identified power potential is half of that) (Onta 2001, p. 107). Revenues from hydropower sales could multiply the growth rate in several Nepalese sectors, including industry, agriculture and tourism (Kayastha 2001, p. 141). But Nepal's current lack of capacity for river diversion or water storage hinders hydropower generation (Condon et al. 2009, p. 13). Nepalese hydropower potential remains largely undeveloped for domestic uses and export. Nepal produces only 714 MW of electricity for its 30 million citizens from all sources of energy, with the result that power outages for several hours a day around the year are common (Chellaney 2011, p. 282). Only 15 per cent of the population has access to electricity and per capita electricity consumption is among the lowest in South Asia (Gyawali 2001, p. 3748). Indeed, a country that could be a major exporter of electricity actually imports power from India (Chellaney 2011, p. 282).

Nepalese hydroelectricity is particularly valuable because it is what is called 'peak power' – electricity that is needed at the time of maximum demand for grid stabilisation. This is important for India because the more inflexible thermal and nuclear power in the north Indian grid, the bigger the need for flexible electricity generation (such as hydropower) to balance the system and prevent its blackout (Gyawali 2013, pp. 7–8). But negotiations over the pricing of peak power have stalled with both camps firmly entrenched in their positions; India refuses to pay more than about three cents per unit, while Nepal feels aggrieved that in return

India proposes selling electricity to Nepal during its hour of load-shedding at double that cost (Gyawali 2013, pp. 5–6).

One major peak power production hydroelectric project is the planned West Seti dam on the confluence of the Seti and Karnali rivers in western Nepal. The dam was originally to be constructed by the Snowy Mountain Engineering Corporation (SMEC) of Australia, but will now be financed by the China Three Gorges Corporation (Hydro World 2015). The West Seti dam will have an installed capacity of 750 MW, most of it for export to India (Onta 2001, p. 114). At 195 metres high the West Seti dam will be one of the world's highest concrete faced rockfill dams, and will impound about 1,500 million cubic metres. The failure of the West Seti dam would have dire consequences for Nepal, resulting in catastrophic loss of life and property (Thapa 2009, p. 11). The Himalayas being a highly seismic region, frequent and high intensity earthquakes pose particular risks to dam infrastructure in the Ganges–Brahmaputra problemshed (Vaidya & Sharma 2014, p. 11).

The development of Nepal's hydropower potential is also partly driven by what Gyawali (2013, pp. 7–8) identifies as the myth that water agreements have to be done in 'hot haste' otherwise Bhutan will capture the Indian power market. Another myth that apparently drives Nepalese hydropower development is that India will develop nuclear power and will no longer need Nepali hydropower. Yet, Bhutan has no Terai plains to irrigate, nor is India's riparian territory adjacent to Bhutan short of water; the state of Assam, which borders on Bhutan, is comparatively water-rich. In contrast, Gyawali points out that Nepal is upstream of the Indian states of Uttar Pradesh and Bihar. These are not only water scarce but happen to be the electoral constituencies of most of independent India's prime ministers and almost half of the Indian parliament. India, argues Gyawali (2013, pp. 7–8), will need storage dams in Nepal for water alone, even if there were no electricity involved. Nepali hydroelectricity can only be a very valuable by-product. How valuable it is exactly is the topic of vigorous debate between state agencies and socio-environmental activists on both sides of the border.

In regard to Bhutan, the kingdom too has not met its potential for hydropower production. Although Bhutan has been developing its hydropower capacity since 1967, only a small fraction of the total estimated potential of 20,000 MW has so far been harnessed (Mukherjee 2001, pp. 149–150). The kingdom is planning to realise half of this identified potential by 2020 (Chellaney 2011, p. 285). Bhutan's existing hydroelectric dam projects have been developed with foreign aid, primarily from India, and India is the largest customer of Bhutanese hydropower (Nexant 2002, p. xiii). India is connected to Bhutanese hydropower through the 336 MW Chukha project, as well as the Kurichu, Chukha Stage II projects, and the 1,000 MW Tala dam (Shah 2001, pp. 28–29). The India-financed Chukha projects are widely regarded as an example of successful bilateral cooperation for mutual benefit, and the revenues from the sale of electricity to India have been so large in proportion to Bhutan's GDP that they have made the country prosperous (Iyer 2007, p. 190).

Hydropower generation: the view from the World Bank

The World Bank (2014a, p. xvi) notes that '[t]he Himalaya has long been seen as holding enormous hydropower potential, adequate to meet domestic energy needs in Nepal (where potential supplies far outstrip potential demand) and provide a significant surplus for trade in the region'. The Ganges Strategic Basin Assessment has taken this narrative as the starting point for a scientific study into the actual hydropower potential of the basin. The Assessment aims to counter with hydrologic and economic evidence the prevailing assumption that there is untold 'blue gold' waiting to be exploited.

The World Bank team asked, is there substantial untapped hydropower potential in the Ganges Basin? The Ganges Strategic Basin Assessment largely confirmed that, yes, there is substantial untapped hydropower potential in the basin. The team found that in Nepal there are more than 40,000 MW of economically feasible hydropower potential, and that less than 2 per cent of this potential has been developed. Furthermore, the 23 largest dams currently under consideration in Nepal would have an installed capacity of about 25,000 MW and could enrich Nepal by some US$5 billion annually – a significant sum relative to Nepal's 2011 GDP of $18.9 billion. The Ganges Strategic Basin Assessment warns, however, that hydropower development on this scale would take many years, require considerable capital investment, and entail effective management of sedimentation. The Assessment concludes that '[n]onetheless, hydropower is an important source of clean energy in a region that is enjoying high economic growth and hence rapidly growing power demands' (World Bank 2014a, p. xvi).

Other issues

Irrigation

Although the issue of hydropower generation seems the most controversial one in the Ganges–Brahmaputra problemshed, it is irrigation that has broader effects in terms of human security, livelihoods and food production. Indeed, India's top two water governance priorities (as listed in the National Water Policy of 1987 and 2002) have been drinking water followed by irrigation. The 2012 National Water Policy did away with the explicit priorities list (Seth 2012), but irrigation remains a fundamental issue for India and the region. As Dinar et al. (2007, p. 254) point out, agriculture accounts for nearly half of all freshwater usage in the Ganges–Brahmaputra–Meghna basin. This makes water supply one of the most significant barriers to economic development (Dinar et al. 2007, p. 254). In India alone, 80 per cent of total water resources are used for irrigation (Kumar & Furlong 2012, p. 7). Irrigation using both groundwater and surface water resources has increased significantly since the Green Revolution of the 1960s and 1970s. But irrigation continues to be a largely inefficient practice that does not on the whole maximise agricultural output and, according to Hill (2013, p. 246), is a contributing factor to the region's multiple water crises.

Irrigation presents a significant challenge for water governance. In India, it is a challenge that has not been met through institutional reform. There is no separate department for irrigation in India. Instead, irrigation falls under each state's Department of Public Works, the primary function of which is the construction of roads and governmental buildings, and providing material requirements and construction of infrastructure for drinking water and irrigation water needs. This results, as Saravanan highlights, in limited attention to water planning and management (Saravanan 2008, pp. 233–235).

The main thrust of policy and planning throughout the twentieth century regarding irrigation was to make more water available for use through what Iyer (2007, pp. 34–35) calls an engineering-dominated supply-side approach that understands water as a commodity. This approach to water, combined with a reliance on irrigated agriculture, has led to India becoming one of the 'big dam' countries and supplying approximately 35 per cent of irrigation water from large storage reservoirs (Bandyopadhyay et al. 2002, p. 4108). Indeed, one reason given by Indian officials for the controversial National River Linking Project 'is the need to increase the country's irrigation capacity in order to meet the growing demand for grain, which, by 2050, would have to be sufficient to feed an estimated 1.5 billion people' (Hangzo 2012, pp. 4–5).

The National River Linking Project, should it go ahead, may alleviate some irrigation water shortfalls for India, but will likely exacerbate water problems for downstream Bangladesh (Condon et al. 2009, p. 11). For Bangladesh, the question of irrigation is one of water quality as well as quantity. As Dinar et al. explain, Bangladeshi aquifers are particularly sensitive to seawater intrusion as 30 per cent of the nation's land is below the high-tide level. During the dry season, intrusion of saline water renders even the little water available in Bangladesh's rivers of poor quality for both drinking and irrigation. Therefore, it is critical that Bangladesh maintains a minimum water flow in certain rivers for salinity control (Dinar et al. 2007, p. 253). India's current and planned upstream water withdrawals endanger this minimum flow into Bangladesh (Dinar et al. 2007, p. 252).

At the same time, there exists scope for Bangladesh to negotiate with India mutual benefits from an inter-basin transfer scheme. Diverting the Brahmaputra water to the Ganges basin could benefit both countries by putting more water in the Padma and the Gorai Rivers in Bangladesh during the dry season (Colopy 2012, p. 320). The transboundary water interaction between India and Bangladesh, however, is currently not amenable to negotiating a benefit sharing arrangement such as this. There is also, according to Colopy (2012, p. 313), a prevailing attitude within the Indian hydrocracy that the water of the Ganges is 'somewhat wasted' on Bangladesh because it suffers from floods even more than others in South Asia, while India is crying out for water. That need, argues Colopy (2012, p. 313), combined with being the upper riparian, serves as justification for keeping the water for upstream use.

For Nepal and Bhutan, irrigation is also an important issue but not one that can readily be addressed by water transfer schemes. As Kayastha notes, diverting rivers is not a complete solution to Nepal's irrigation problems, meaning that it will have to develop its groundwater resources to fulfil irrigation needs (Kayastha 2001, p. 123). As for Bhutan, the agricultural sector consumes the highest percentage of water (Royal Government of Bhutan 2003, p. 8) although a mere 8.8 per cent of the land in Bhutan is under agricultural use, and only 25 per cent of cultivatable land is under irrigation (Mukherjee 2001, p. 149). As the Bhutanese population grows and demand for water increases, the Royal Bhutanese Government is concerned about increased efficiency ('more crop per drop') through adaptive and applied research, rather than increasing the availability of water for irrigation (Royal Government of Bhutan 2003, p. 8). Nonetheless, despite these attempts at improving irrigation practices through means other than large-scale infrastructure, engineering-dominated supply-side approaches dominate the management of water for irrigation throughout the Ganges–Brahmaputra problemshed.

Flooding

Much like the issue of irrigation, the issue of flooding hinges largely on the prevailing paradigm of 'command and control'. One illustration of this is the devastating floods that took place in India's northern Bihar province in 2008. The flooding, argues Colopy (2012, p. 215), was framed as a fault of nature, rather than a fault of the 'command and control' approach to water management. Yet it was this mentality that led to the construction of numerous embankments along the naturally unpredictable Kosi River, which flows from Nepal into northern India and into the Ganges. The failure of these embankments exacerbated the devastation of the flood (Colopy 2012, p. 215).

Upward of two thousand miles of embankments had been built to control the annual monsoonal floods that briefly cover the plains of Bihar and enrich the soils with sediment for post-monsoon crops. The consequences of the embankments, however, have been the opposite of that which was intended. As Colopy (2012, p. 215) points out, where there were six million acres of flood-prone area in 1952, now there are seventeen million. She argues that where there was inconvenience there is now a yearly disaster. Those living on the inside of the embankments now face the risk of a much higher flood than previously when the water covered a larger area but in a thinner layer. Meanwhile, those living on the outside have been pitted against them because of the vested interest they now have in maintaining the system of embankments, which also serve as roads (Colopy 2012, pp. 224–225).

The problem of the embankments is not that they overflow, but rather that they are often poorly maintained and fail. The Ganges Strategic Basin Assessment shows that most flooding events in the basin are caused by embankment failures, high flows in small tributaries, and localised rainfall. The study finds that lowering flood peaks within embanked rivers through dam storage is unlikely to have a significant effect on flooding events (World Bank 2014a, pp. xiv–xv). Asthana and Shukla (2014, p. 63) show that the flood-control strategy using dams has been only moderately successful in practice. Yet, the 'command and control' school of thought continues, according to Colopy (2012, pp. 215–216), to dominate responses to flooding in this region.

The Ganges Strategic Basin Assessment attempts to reframe the discourse as one of flood management rather than flood control. The study asks 'is large infrastructure the best strategy for protecting communities from floods?' and finds that, contrary to popular belief, a strategy exclusively focused on large infrastructure cannot protect basin communities. There is, the authors argue, no simple solution to the problem of flooding on the Ganges plains because the highly variable monsoon-driven Ganges system with its thousands of tributaries cannot be effectively controlled with a one-size-fits-all approach. Instead, flood protection for basin communities and the livelihoods of their people requires a broad, balanced combination of 'hard' and 'soft', as well as local and transboundary, responses. The World Bank proposes that 'a shift in focus is needed from "flood control" to "flood management"' marked by: greater emphasis on regional forecasting and warning systems; embankment asset management; drainage; disaster preparedness; land use zoning; safe havens; flood insurance; and training and communications campaigns (World Bank 2014a, p. xvii).

Floods are a natural part of the hydrology of the Ganges–Brahmaputra problemshed and serve essential ecosystem functions such as groundwater recharge, soil fertilisation and fish stock renewal (Thomas 2012). Flooding also brings nutrient-rich sediment. At the same time, flooding is destructive, but agricultural practices have been adapted to safeguard against some flood risks. For example, rice harvests were segmented in pre-colonial times throughout the year to hedge against flood pattern fluctuations. The understanding of floods as bringing benefits as well as danger was once part of water management. But this understanding, which is fundamental to flood risk management, has been lost with the development of the 'command and control' approach to water governance (D'Souza 2007, pp. 263–264).

The mental shift from flood control to management presents its own challenges for water governance, but may, counter-intuitively, simplify one aspect of transboundary interactions in the Ganges–Brahmaputra problemshed. Dr Claudia Sadoff, one of the authors of the Ganges Strategic Basin Assessment, has said that '[i]f you can't reduce floods in Bihar or saline intrusion in the Sunderban with infrastructure in Nepal, then that simplifies the discussion. Maybe you don't have to negotiate everything across boundaries' (Sadoff cited in Colopy 2012, p. 320). Flood management, then, becomes largely a local issue (even in border areas) rather than a transboundary one because it cannot be effectively managed across the entire basin. Taking a problemshed view, rather than a watershed one, allows for reconceptualising transboundary water issues according to these social, economic and management realities rather than constricting them to strictly hydrological boundaries.

In the Ganges–Brahmaputra problemshed, Bangladesh is the country most prone to floods, but India and Nepal also face flooding risk during the wet season

(Dinar et al. 2007, p. 252). If, as the Ganges Strategic Basin Assessment shows, all these flood events cannot be controlled or mitigated through upstream water storage infrastructure in Nepal, then the issue need not remain on the transboundary water governance agenda except insofar as mutual assistance and the sharing of best practice are concerned.

Sedimentation

The problem of flooding, especially for Bangladesh, is not just one of a surfeit of water, but also a surfeit of sedimentation. Compounding the diversions of Ganges River flow to Bangladesh is the rapid rise in silt deposition rates since the Farakka Barrage was erected upstream in India in 1975. Such sedimentation substantially reduces the river bed's storage capacity and further increases flood risks (Thomas 2012). This is another example of how environmental processes coincide and create natural disasters in the Ganges–Brahmaputra problemshed (National Research Council of the National Academies 2012, p. 82). Because of the risks posed by high levels of sedimentation, the Ganges Strategic Basin Assessment asks '[i]s it possible to control sediment in the Ganges?' (World Bank 2014a, p. 49).

There is a belief within the Ganges basin that, as elsewhere in the world, a combination of erosion management and upstream storage structures might control sedimentation in the river. According to the Ganges Strategic Basin Assessment, however, this basin is different. Erosion from the Himalayan mountains makes the Ganges one of the three most sediment-laden rivers in the world. Both the high volume and the source of this sediment make it extremely difficult to manage. The volume of sediment is so large that capturing it behind large dams would be extremely costly. The Ganges Strategic Basin Assessment has found that the reservoirs behind these large, expensive structures would fill quickly and, thereafter, produce very few benefits. Moreover, the high altitude and terrain of the sediment source regions, as well as the nature of the sediment and the ongoing tectonic processes, pose significant risks to the viability of infrastructure on the scale necessary to have any measurable impact on basin sediment loads (World Bank 2014a, pp. xvii–xviii).

The problem of sedimentation cannot be readily resolved with infrastructure and water storage. Indeed, sedimentation may be, in the words of analysts from ICIMOD, 'the greatest challenge for both large and small surface water storage reservoirs in the region' and the problem is likely to increase with climate change (Vaidya & Sharma 2014, pp. 11, 31). It is also currently unclear to what extent dams in Nepal can efficiently pass the large amounts of sediment eroded from high in the Himalaya (World Bank 2014a, p. xiv).

The fact that sedimentation endangers dams and that dams cannot be effectively used to mitigate the effects of sedimentation downstream are both questions of control. As with the issue of flooding, however, the question of control misses the most important point for water governance. As the authors of the Ganges Strategic Basin Assessment argue, '[s]ediment, like floods, is a challenge that must be managed in the Ganges; it cannot be fully controlled' (World Bank 2014a, pp. xvii–xviii). The prevailing water paradigm implies that rivers – including the sediment they carry – must be controlled, and that they must be controlled through human engineering. Instead, a management approach may prove more fruitful.

Sedimentation is not necessarily undesirable; indeed, it brings many benefits. High-silt floods fertilise soils, for example (Thomas 2012). In the pre-colonial era, floods on the Gangetic plain had been understood as a duality of calamity and benefits (D'Souza 2007, 264–265). Even the devastating 2008 flood in Bihar state brought with it nutritious sediment that enriched the soil (Colopy 2012, pp. 215–216).

High sedimentation and flooding are a mixed blessing, but under the 'command and control' paradigm they are both seen almost exclusively as a problem. This limits the potential for developing innovative, positive-sum solutions. There needs to be a return to the non-binary understanding of sedimentation and flooding, that is, an understanding that each brings both advantages and disadvantages (similarly to the nuanced understanding of water conflict and cooperation coexisting, with conflict sometimes having positive outcomes and ostensible cooperation leading to negative ones). This is both a paradigm shift desired by riparian stakeholders (as discussed in Chapter 5) and one that international actors are, in part, attempting to institute already (as discussed in Chapter 6).

4 Why are international actors interested in the Ganges–Brahmaputra problemshed?

Why have the hydropolitics in South Asia attracted the attention and funding of foreign aid donors in the past decade? Chapters 2 and 3 illustrated that there are many unresolved issues on the transboundary water agenda in the Ganges–Brahmaputra problemshed. International organisations and foreign aid donors, however, are driven by more than just the needs of the problemshed – they are also driven by global dynamics and their own agendas.

This chapter examines the global context in which these international actors operate, and how this affects the choice to intervene in the Ganges–Brahmaputra problemshed. For a start, there is increasing international interest in South Asia generally. The region is a potential hotspot of instability, even conflict, arising out of environmental risks, including water scarcity and water insecurity. At the same time, the World Bank is no longer the main financier of water infrastructure (e.g., dams) in the region. It is attempting to rebuild its influence there while facing increasing competition from new funding agencies, such as the Asia Infrastructure Investment Bank (AIIB), and India's growing wealth. India is no longer a net aid recipient and that has led to a shift in aid strategies from donors such as Australia and the United Kingdom. These are now less focused on direct bilateral aid and more focused on regional partnerships in collective action problems, such as water governance. The general attitude towards dams and dam-building has also shifted since the 1990s, and that too has influenced what international organisations and foreign aid donors are trying to achieve in the Ganges–Brahmaputra problemshed in regard to transboundary water governance.

Increasing international interest in South Asia

South Asia is increasingly recognised as particularly vulnerable to climatic and demographic shocks. The region is thus a potential hotspot of instability, even conflict, arising out of environmental risks, including water scarcity and water insecurity. Recent literature acknowledges that 'in the next several decades, South Asia may be the linchpin of humanity's collective effort' on non-traditional security threats (Matthew 2014, p. x). Water scholars are more and more likely to believe that 'hydropolitics between India and its neighbouring riparian states is one of the most urgent, complex, and contentious issues facing the regional community in the twenty-first century' (Asthana & Shukla 2014, p. 241). This represents a shift in scholarship on transboundary waters.

Until recently, scholars of transboundary water interaction have tended to study basins with the potential for 'real' conflict (e.g., the Jordan, Nile, Tigris-Euphrates and in South Asia, the Indus) rather than those with potential for non-traditional conflicts. As one regional analyst argues, there is still an 'Indus bias' among the scholars and aid donors in South Asia. Much of the foreign aid money goes towards the Indus basin because it is considered more fragile, even though the tensions over the Indus are not a geopolitical threat to India and Pakistan because the basin is governed by the Indus Waters Treaty, a relatively stable mechanism (Prasai 2014). Hill (2015, p. 740), too, notes that in recent years many think tanks have been collaborating with external agencies to promote dialogue in the Indus basin rather than the eastern Himalayas. This 'Indus bias', however, may slowly disappear as non-traditional security threats, rather than the threat of state-on-state warfare, become increasingly important to international organisations and aid donors.

The Ganges–Brahmaputra problemshed presents a perfect storm for non-traditional security threats and collective action problems: socio-political instability (National Research Council of the National Academies 2012, p. 84); rapidly rising populations (Wirsing et al. 2013, p. 14); increasing demand for food, energy and water (Chellaney 2014, p. 622); high poverty rates (Shah 2001, pp. 18–22); ecological degradation (Wohl 2012, p. 3); and, of course, the most pressing of contemporary global concerns, climate change (Vaidya & Sharma 2014, p. 3). The World Bank envisions that climate change will have 'cascading impacts' on South Asia and predicts extreme monsoon variability, increases in mortality rates, regional crop production failing to meet food demand, leading to malnutrition and stunting for generations to come (Evans 2015b, p. 80). These bleak scenarios are part of the reason that the World Bank is investing heavily in climate change resilience strategies and engaging in water security and transboundary river management issues in South Asia (Evans 2015a, 2015b, pp. 64, 80).

Another aspect to increased engagement on non-traditional security threats is that they incorporate the nexus of food, water and food security. As such, argues Hill (2015, p. 740), integration of upstream and downstream communities in the Himalayas must be done in a way that accounts for the many ecosystems of the region. This, in turn, provides opportunities for international engagement and capacity building.

Waning influence of the World Bank in South Asia

The collective action problems, such as transboundary water conflicts, facing South Asia are of concern to foreign aid donors, but they are not the only driver of international actors intervening here. Geopolitics is also a factor. In recent decades the World Bank's influence in South Asia, especially in regard to large infrastructure projects such as hydropower dams, has been waning. This may be in part why the World Bank has been shifting towards capacity building and governance projects in this region. Earle et al. argue that the World Bank had also hoped to replicate in South Asia the common water-sharing framework of the Nile Basin Initiative (Earle et al. 2015, p. 41). As for the other significant aid donors engaging in transboundary water governance in the region – Australia, Japan, Norway and the UK – these states have been decreasing their bilateral aid spending to South Asian countries, and to India in particular. As such, they are finding new ways of retaining presence, influence and relevance in the region.

The World Bank has historically been influential in setting the water governance agenda in developing countries. Initially, this influence was exerted through financial mechanisms and incentives; later it came to be ideational. Stein shows that in the first decades of its existence, the World Bank focused on structural adjustment of developing economies, liberalisation of markets, and privatisation of state assets on the assumption that private property ownership would encourage greater efficiency, investment and growth. But by 1989 the World Bank began to expand its developmental lexicon. New initiatives in developing countries suddenly focused on issues such as governance and capacity building, social capital, poverty reduction, sustainable development, decentralisation and country-level ownership of aid policies (Stein 2008, p. 26). Both of these approaches – financial assistance and capacity building – have facilitated the World Bank's water governance agenda, one predicated on the commodification of natural resources.

The World Bank has its genesis in the disintegration of the colonial world order and the paradigm shift from colonial economics to development economics (Piper 2014, pp. 20–22). As Piper (2014, p. 20) argues, the creation of the World Bank also signified a shift in global power from European to American interests following World War II. Since then, the World Bank has been a primary proponent of the commodity approach to water, which supports the idea that economic and market tools are to be used to ensure the efficient provision of water (Langford 2005, p. 274). Poor countries are thus pushed into privatisation of water facilities by agencies such the World Bank, which provide incentives through foreign aid and loan packages (Elver 2006, p. 894).

The financial incentives of the World Bank, as Homer-Dixon (1995, p. 588) argues, inform commentary in influential business-orientated newspapers, magazines and books, and thus have become the dominant paradigm of water governance. According to Allan (2006, p. 48), this represents the fourth wave of five water management paradigms. This fourth wave was:

> inspired by economists who had drawn the attention of water users in the North to the economic value of water and its importance as a scarce economic input. These ideas gained currency in the early 1990s. There has been an attempt to export them to the South via agencies such as the World Bank.

This affirms Sneddon's claim (2013, p. 16) that hegemonic concepts are tangibly connected to mechanisms of world hegemony such as international organisations which legitimise certain norms that serve specific interests or co-opt the elites of dependent countries. As such, Piper (2014, pp. 20–22) argues that the World Bank is propagating colonial attitudes and that these are manifest in its policies on water.

In South Asia, the World Bank has historically applied the neo-liberal rationale to dam building. By the late 1980s, the World Bank was the world's biggest financier of big dams, but also found it difficult to promote big dams in light of rising protests against its policies (Goodland 2010, p. 384). As Goodland (2010, p. 384) illustrates, 'from the 1970s to the mid-1990s, the World Bank created a monumental imbroglio of the entire hydroelectric dam sector, and worse, denied there were any shambles; and even if there were, the [World Bank Group] was not responsible for them'. Several World Bank funded large dam projects in India in particular prompted in the early 2000s a significant review of the Bank's dam-building worldwide. These reviews undermined the World Bank's position in South Asia as the main financier of large dams and also shifted international attitudes toward dams generally. The advent of China as a major financier of large dams (for example in the Mekong and Nile basins) has also challenged the World Bank's near-monopoly on funding such infrastructure projects (Earle et al. 2015, p. 42). These factors contribute to the World Bank interest in improving water governance in South Asia.

Changing public attitudes toward dam building

The waning public support for dams around the world, but especially in India, has also shaped the World Bank's stance on transboundary water governance in the Ganges–Brahmaputra problemshed. Large dams enjoyed their heyday in the twentieth century, but the tide of popularity turned against them around the year 2000 with the World Commission on Dams. Since then, dams have been slowly returning to the agendas of international funding organisations (including the World Bank), albeit not all dams; large and mega dams are still out of favour, but multi-purpose and micro-hydro projects are on the rise. This is the case in the Ganges–Brahmaputra problemshed too. The World Bank has been, for example, carefully promoting multi-purpose projects through the Ganges Strategic Basin Assessment (GSBA).

The World Commission on Dams, claim Asthana and Shukla (2014, p. 69), was in part a response to the popular movement against dams in India. One of the most controversial projects was the Tehri dam, a large dam on a northern tributary of the Ganges in India's Uttarakhand province, close to the Chinese border. Opposition to the Tehri dam began in the 1970s and the anti-Tehri movement, argues Sharma (2012, p. 19), has been a significant marker for Indian environmentalism in general and struggles against big dams in particular, especially against the social dislocation that large dams in densely populated areas may cause. The protests in India brought these issues to international attention, and in 2000, the World Commission on Dams found that large dams had displaced 40 to 80 million people worldwide (Piper 2014, pp. 20–22).

At first glance the Commission's findings may seem like a win for the anti-dam movement (which Briscoe (2010, pp. 405–406) argues strongly influenced – even hijacked – the World Commission on Dams). But the Commission was not wholly anti-dam: its stance can be summed up not as 'no dams', but rather 'no bad dams' (Dixit & Gyawali 2010, p. 108). The Commission's final report, *Dams and Development: A New Framework For Decision-Making*, finds that '[d]ams have made an important and significant contribution to human development, and the benefits derived from them have been considerable', but that '[i]n too many cases an unacceptable and often unnecessary price has been paid to secure those benefits, especially in social and environmental terms, by people displaced, by communities downstream, by taxpayers and by the natural environment' (World Commission on Dams 2000, p. xxviii).

The international focus in the 1990s on the 'unnecessary price' of dams had unintended consequences. The anti-dam sentiment was precipitated by what Briscoe (2010, pp. 402–403) calls 'red-green coalitions' (a 'combustible mixture of rich-country anti-capitalists and environmentalists'). It resulted, he claims, in the baby being thrown out with the bathwater: numerous small, locally unopposed dam projects in developing countries were stalled or otherwise undermined by opposition in the global North and hesitation within the World Bank to fund potentially controversial projects (Briscoe 2010, pp. 402–403). Finding a way to maximise the benefits of dams while reducing the toll on people and planet was the primary purpose of the World Commission of Dams.

Briscoe (2010, p. 403) describes the World Commission on Dams as an attempt to bring stakeholders from all sides of the debate together, to develop a shared fact base and a shared set of recommendations. This would ensure that 'bad dams' were not financed or built, and that 'good dams' were financed and built without inordinate costs or delays (Briscoe 2010, p. 403). The result was what Goodland (2010, p. 386) calls 'the most global, holistic, systematic, comprehensive, participatory, and scientifically valid assessment of large dam building to date'. The *Dams and Development* report challenged World Bank practices and established guidelines for the selection of new dam projects. But these were not eagerly adopted. As Goodland (2010, p. 384) shows, 'the Bank, along with the hydropower industry, resisted the guidelines and struggled with reformers' (although, as Briscoe (2010, p. 405) points out, many of these guidelines were onerous, impractical, and even undesirable).

The World Bank was not the only major opponent of the World Commission on Dams report. The Indian government was also initially resistant to the findings of the Commission. Iyer (2001, p. 2275) shows that the report was seen as a conspiracy against the complex and controversial Sardar Sarovar Project on the Narmada River, and the Government of Gujarat in particular lobbied the Indian central government to reject the Commission's 'anti-dam' findings. He also argues that 'Developing countries see it [the *Dams and Development* report] as yet another instance of the imposition on them by the developed countries of an agenda designed in the latter's interests' (Iyer cited in Briscoe 2010, p. 408). Be that as it may, Wood (2007, pp. 145–146) shows that ultimately the World Commission on Dams' findings and the popular movement against the Sardar Sarovar dams did affect both India's water policy and the World Bank's dealings in India. For instance, the Ganges Strategic Basin Assessment produced by a World Bank-led team goes to great lengths to stress 'no bad dams' should be built in the Ganges basin.

The protests against Sardar Sarovar and other dams on the Narmada River in India began in the 1980s and 1990s (Wood 2007, pp. 145–146). As Wood (2007, p. 43) argues, the human rights and ecological issues raised by the anti-dam movement also formed part of a larger theoretical critique of modernisation, which called into question many of the assumptions and value-orientations of either market-driven or state-driven efforts at 'development' (this is a critique of modernisation also largely favoured by Briscoe's (2010, p. 402) 'red-green coalitions'). Wood (2007, p. 195) adds that the public concern over these issues, both within India and internationally, and the protest against dams on the Narmada, led to a re-examination of India's water policy and an exploration of alternative means of developing the country's water resources. The lasting legacy of the Sardar Sarovar controversy, according to Wood (2007, p. 195), is that large dams are seen not to live up to expectations in terms of cost, construction time and efficiency of output.

The Sardar Sarovar controversy left a legacy on the World Bank too. The Bank conducted an independent review of the beleaguered project (Wood 2007, p. 167) and consequently changed its policies on dams, resettlement and rehabilitation and the monitoring and assessment of development projects (Whitehead 2008, p. 296). These new policies are evident in the way the World Bank has scoped and conducted the South Asia Water Initiative (SAWI), focusing mostly on governance and capacity building.

Shifting focus of the World Bank away from infrastructure and toward governance

With the unpopularity of large dams, and growing complications of funding such projects, the World Bank began to turn towards 'softer' areas of development aid, such as capacity building and governance. In South Asia, this transition echoes the World Bank's first major foray into the region, when in the early 1960s it was still the Bank for Reconstruction and Development and mediated the negotiations over the Indus Waters Treaty. The Bank's role was initially to provide good offices (i.e., mediation services) for India and Pakistan, but over time evolved into conciliation between the two parties, actively finding a solution and ultimately providing significant financial assistance as a condition of their consent to the terms of settlement (Cosgrove 2003, p. 39).

In recent years, the World Bank has also taken a leadership role in a number of large-scale partnerships designed to improve the delivery of global public goods – such as knowledge, water and sanitation, and education – for developing countries (Evans et al. 2015, pp. 189–190). Thus, what the World Bank is doing has changed, and so has the policy of collaboration. As Evans et al. show, in the past 15 years, partly as a result of the Millennium Development Goals (MDG), the World Bank has dramatically increased engagement with a broader range of stakeholders, especially those from civil society organisations and sub-state actors. While such collaboration is dynamic and innovative, it is also a challenge for the World Bank, which was designed to work on an international scale to engage with sovereign nations (Evans et al. 2015, pp. 189–190). The Abu Dhabi Dialogue (discussed in Chapter 6) is an example of this type of engagement with a broad range of state and non-state stakeholders.

The largest investment the World Bank has made in recent years in a collaboratively governed, non-infrastructure related enterprise has been in the pursuit of climate change related goals, and climate change comprises the largest share of the World Bank's environment portfolio (Evans 2015a, p. 64). This emphasis on climate change adaptation has had the effect, as Smith argues, of bringing dams and water infrastructure back up the development financing agenda (Smith 2010, p. 441). In part this is because hydropower is sometimes considered to be renewable energy, and in part because the World Bank is, according to Briscoe (2010, p. 411), 're-embracing infrastructure in general and water infrastructure in particular'.

The type of infrastructure creeping up the international development agenda is not the type of infrastructure that was previously there. The current emphasis is on good governance through multi-purpose projects, and the delivery of multiple benefits. In recent World Bank documents, proposed dam projects are touted for their ostensible function of flood control, water storage for irrigation and so on. One example of this is the Ganges Strategic Basin Assessment (GSBA), which concludes that on the Ganges basin, upstream water storage (that is, man-made reservoirs) can augment dry season water flows downstream, and that there is significant untapped hydropower potential that could bring big benefits and few trade-offs to the riparians in the Ganges basin (World Bank 2014a, pp. 51–64, 79–86). At the same time, the report is careful to point out that the benefits of multi-purpose dams – such as flood control – do not exist in a vacuum, and instead can only be realised in conjunction with robust governance and 'soft' intervention (World Bank 2014a, pp. 81–86). As such, the Ganges Strategic Basin Assessment embodies the World Bank's evolving position on large infrastructure projects, governance and the provision of global public goods.

This change in operational focus explains *what* the World Bank is currently trying to do in South Asia, but *why* it is there at all is best explained by strategy. The World Bank had been instrumental in negotiating and funding the Indus Waters Treaty and then in funding of major infrastructure projects in South Asia. Now, as one Indian water analyst pointed out, there are other sources of such funds, such as new regional development banks (Ray 2014). Indeed, whereas the World Bank finances 'a handful of dams around the developing world, China now finances over 200 outside of China' (Briscoe 2010, p. 412). China is also increasingly using foreign aid as a powerful tool of statecraft (Brant 2014). Moreover, most new infrastructure projects, especially in India, are privately funded, and so the traditional role of the World Bank as financier of infrastructure is quickly diminishing (Ray 2014). This too in part explains the shift to a new role in governance, capacity building, and the two-for-the-price-of-one sale of multi-purpose projects.

The World Bank is operating in an increasingly competitive market in South Asia. The emergence of the New Development Bank (a.k.a. the BRICS Bank) (VI BRICS Summit 2014) and the Chinese-led Asia Infrastructure Investment Bank (AIIB) (Chapman 2014) brings two competitors for the World Bank into the region. With its hegemony thus threatened, and the countries of South Asia – India in particular – less reliant on its funding, the World Bank is forced to redefine its role in the region if it is to stay relevant, effective and influential there.

The dynamics of foreign aid in South Asia, both in terms of donor objectives and recipient priorities, are changing too. Most significantly, India is shifting from being a net recipient of foreign aid to being a regional aid donor in its own right and this is affecting the strategies of previously significant donors to India. In the past decade, India has been transitioning to being an 'emerging donor' country, and as part of its 'South–South' strategy, 'India seeks to use its foreign aid programs as a tool to further its own economic, political, and strategic interests while also taking into account the needs of the recipient country' (Library of Congress 2015). Accommodating the needs of neighbours may also support India's interest in emerging as a regional, and even global, leader. As Jones (2008, pp. 183–193) argues, India can only achieve its goals on the world stage by operating beyond South Asia. But this requires India to take a lead role in stabilising its own region as part of being accepted by other great powers as a member of the club. This may be why India ultimately has been participating in the World Bank-led Abu Dhabi Dialogue despite, as discussed in Chapter 7, being reluctant to do so.

Jones was writing about India's global leadership ambitions in 2008, but his words are just as applicable, if not more so, to Narendra Modi's prime ministership of India commencing in 2014. During Modi's term in office, India has been continuing to increase regional leadership while reducing its dependence on foreign aid. India's net foreign aid receipts have fallen sharply each year since 2010–11, and will likely continue to drop (DEVEX 2014). India no longer accepts bilateral aid from the main international actors in water governance in the region, Australia (DFAT n.d.) and the United Kingdom (*Times of India* 2012), and this has fundamentally shifted the way in which these countries project and maintain influence in South Asia. Focusing on transboundary water governance is a way for international actors to continue buy-in and presence in the region without impinging on India's emerging strategy. Water resource management is a convenient policy niche for these actors, not only because it aligns with the prevailing values of the international development agenda (climate change resilience, good governance, sustainable development etc.) but also because it is not (yet?) a priority for India.

In conclusion, the commercial, geopolitical and strategic interests of international actors are all drivers of interventions in the transboundary water governance of the Ganges–Brahmaputra. These interests are as significant in motivating them as is the interest in increasing water cooperation between India, Nepal, Bhutan and Bangladesh.

5 Normative claims of international actors

International organisations and foreign aid donors engaging in transboundary water interaction in the Ganges–Brahmaputra problemshed do not do so in an intellectual or ideological vacuum. They are instead the product and the propagators of numerous narratives, ideas and ideals. These values are also the basis of the normative claims that international organisations and foreign aid donors make in regard to transboundary water governance. The global paradigms and discourses discussed below are used, in part, to justify interventions in the Ganges–Brahmaputra problemshed, and to guide their aims here.

This chapter discusses the paradigms and discourses that factor into the justifications for foreign-led interventions in transboundary water governance. For a start, the primacy of good governance has become a value in many areas of international development, including water management. Good water governance espouses the principles of participation, deliberation, consensus-building and transparency. Meanwhile, the rise of non-traditional security threats and collective action problems such as climate change has increased global attention on environmental risks in volatile parts of the world, including South Asia. Water security is also an increasingly important challenge for development agencies, especially considering the nexus with food and energy security. At the same time, the emergence of the sustainable development agenda (made explicit in 2015 with the United Nations-sanctioned Sustainable Development Goals) has changed the way in which international donors approach environmental factors in developing countries. One example of this is the beginnings of a shift away from a 'command and control' approach to one based on ecological principles. Each of these narratives and normative claims is discussed in turn.

The ideas and values examined in this chapter can be categorised as what Sneddon (2013, p. 21) calls 'hegemonic concepts'. He argues that understanding hegemonic concepts illustrates how seemingly 'global' ideas hatched in the forums of powerful transnational actors and organisations become applied across special scales (international, national, local). This is of particular importance in the case of the Ganges–Brahmaputra problemshed; the hegemonic concepts of good water governance, sustainable development and so on are the basis of normative claims made by international actors intervening in transboundary water governance in this problemshed. Understanding the ideas and values that dominate discourses on transboundary water cooperation is important not only because they drive international actors into the Ganges–Brahmaputra problemshed, and affect the scope of their interventions there. As Sneddon (2013, p. 21) shows, understanding hegemonic concepts – such as those of good governance and sustainable development – also allows for an exploration of counter-hegemonic alternatives to water governance.

Good water governance

As Nishat (2014, p. 110) shows, the principles of water management being applied in the Ganges–Brahmaputra problemshed are not necessarily those that come from within the region, but rather are those driven by multinational institutions such as UN Water, UNDP, FAO and international civil society organisations such as the Global Water Partnership and the World Water Council. Water policy discussion within the Ganges–Brahmaputra problemshed can therefore be said to be generally influenced by overarching global discourses.

One such global discourse is, according to Routray (2010, p. 33), that of Integrated Water Resource Management (IWRM). Kurup (2010, p. 157) shows that in the last decade in particular, an increased emphasis in South Asia has been on the understanding of the management of water resources using the IWRM paradigm, and this is due directly to its adaptation and endorsement by powerful international agencies like the Global Water Partnership and the United Nations.

The recent focus on what Mollinga and Gondhalekar (2014, p. 180) call the 'water question' is in itself a value-laden prospect. Mollinga and Gondhalekar argue that since the early 1990s, water resources management has become globalised and aligned with the 'three big ideas' of the 'age of environmental critiques' that dawned around the 1992 Rio Earth Summit. These three big ideas are: the market, democracy (often phrased as 'good governance') and sustainability. These ideas were assembled in the concept of Integrated Water Resources Management (IWRM). Since around the year 2000, this 'new paradigm' of water governance has been actively propagated through international development funding agencies and knowledge organisations involved in capacity building (Mollinga & Gondhalekar 2014, p. 180). International actors in the Ganges–Brahmaputra problemshed can also be said to be part of this epistemic community propagating this 'sanctioned discourse'.

The paradigm of water resource management that espouses the values of the free market, good governance and sustainable development has come into its own in the era of climate change and other non-traditional security threats. As Mollinga and Gondhalekar (2014, p. 180) point out, '[t]he recent surge in attention to climate change, in debates on which water resources play a central role, has added to the perceived global nature of the "water question", most recently popularly framed as a "water security" question'. Indeed, water security is emerging as a discrete policy area, with distinct applications for both national security and human development. Water is also increasingly prominent on the international development agenda, as evidenced by the Sustainable Development Goals (SDG) focusing on water issues other than just sanitation and health. Meanwhile, the international development agenda is increasingly closely aligned with the international environmental conservation agenda – again, as evidenced by the growing policy focus on sustainable development. Throughout these interlinked discourses the recurring theme is that of governance; bad or weak governance is at the root of many contemporary global challenges, while good governance is touted as a necessary factor in addressing them.

The concept of good water governance is now well entrenched in the scholarship on and praxis of water resources. According to UN Water, it is critical to Goal 6, the water goal, of the SDGs: good governance 'underpins all the water targets and supports linkages to other development themes' (UN Water 2014). But it was not always at the forefront of water discourse.

The concept of good water governance has evolved from the relatively recent understanding of water as being inherently political. As Mollinga (2008b, pp. 8–9) points out, as recently as the late 1990s, 'politics and the political were anathema in most water policy circles. The social engineering paradigm reigned largely unquestioned.' He goes on to argue that the rise of the theme of good governance generally brought politics into the discourse of water resources development (Mollinga 2008b, pp. 8–9). At the same time, this discourse was being broadened to include aspects of human development and environmental conservation.

The role of water management in reconciling human development with environmental considerations has its genesis in the 1992 International Conference on Water and the Environment. According to Muller (2015, p. 675), the so-called Dublin Principles, which were an outcome of the conference, were radically new forms of deliberative democracy, environmental governance and multilateralism in the water sector globally.

According to the United Nations (2009, p. 28), good governance 'is participatory, consensus oriented, accountable, transparent, responsive, equitable and inclusive, effective and efficient and follows the rule of law'. That is simultaneously a narrow definition, but also one that remains open to interpretation. As Sachs (2015, p. 42) emphasises, good governance is many different things. He explains that both the public sector (government) and the private sector (business) should operate according to the rule of law, with accountability, transparency and responsiveness to the needs of stakeholders. Active engagement of the public on critical issues such as land use, pollution, and the fairness and honesty of political and business practices is another characteristic of good governance. Sachs does not refer to the role of civil society in his definition of good governance, but the inclusion in governance processes of actors that are neither government nor business is also implicit in the understanding of good governance.

In 1987 the World Commission on Environment and Development in its seminal report *Our Common Future* (a.k.a. the Brundtland Report) called for the collaboration of various actor groups in the good governance of environmental resources: 'Environmental protection and sustainable development must be an integral part of the mandates of all agencies of governments, of international organizations, and of major private sector institutions' (1987, Chapter 12, Section I, Paragraph 1.17). Similar sentiments were echoed in 2012, at the Rio+20 conference. *The Future We Want*, the outcome document of those proceedings, noted:

> the valuable contributions that non-governmental organizations could and do make in promoting sustainable development through their well-established and diverse experience, expertise and capacity, especially in the area of analysis, sharing of information and knowledge, promotion of dialogue and support of implementation of sustainable development.
>
> (UN 2012, paragraph 53)

The same report acknowledges the importance of multi-level cooperation: 'We recognize that effective governance at local, sub-national, national, regional and global levels representing the voices and interests of all is critical for advancing sustainable development' (UN 2012, paragraph 76).

This emphasis on inclusion is also a fundamental aspect of deliberative governance – a method of good governance. Deliberative governance is, according to Dore (2014, p. 195), a constructive engagement between stakeholders over issues of mutual interest, such as water resource. This type of good governance promotes 'inclusive, deliberative processes that emphasise different perspectives, critical analysis, learning and institution-building whilst respecting rights, accounting for risks, acknowledging responsibilities and fairly distributing rewards' (Dore 2014, p. 195). Deliberative governance can be applied to transboundary water issues, and indeed this is the basis of the main foreign-led interventions in the Ganges–Brahmaputra problemshed.

An important element of deliberative governance specifically, and good governance more generally, is that of communication and information sharing. The Brundtland Report observes that '[c]ooperation on environmental issues among developing countries has often been made difficult by poor communications' (World Commission on Environment and Development 1987, Chapter 11, Section III, Paragraph 2.41). The Rio+20 outcome document also notes that access to information is essential to the promotion of sustainable development (UN 2012, paragraph 43). It resolves to 'promote the science-policy interface through inclusive, evidence-based and transparent scientific assessments, as well as access to reliable, relevant and timely data in areas related to the three dimensions of sustainable development, building on existing mechanisms, as appropriate' (UN 2012, paragraph 76(g)).

In terms of water management, evidence-led policy and knowledge sharing is particularly important: technical and scientific cooperation is a pillar of good water governance. As Gyawali and Schwank (cited in Chokkakula 2012, p. 37) make clear, water sharing agreements based on a good scientific understanding by both the riparians of the river system, of its behaviour and capabilities, will lead to peace and understanding that last. Meanwhile, those with scientific uncertainties will always provide opportunities for strife and disputes. But as discussed in Chapter 8, there is a culture of secrecy around water-related data in the

Ganges–Brahmaputra problemshed. Furthermore, as Dinar et al. (2007, p. 250) point out, it is difficult if not impossible to find accurate hydrological data on the Ganges–Brahmaputra–Meghna mega-basin in part because the system of rivers is so complex it is difficult to disaggregate how much water comes from where.

The World Bank is particularly aware of the limited technical and scientific cooperation in the Ganges–Brahmaputra problemshed, and the difficulty of accessing accurate hydrological data. As the Ganges Strategic Basin Assessment makes clear, 'there has been no basin-wide knowledge base and analytical framework that could be used by riparian states to explore options and facilitate cooperative planning in the Ganges. Information and data are surprisingly scarce and difficult to obtain' (World Bank 2014a, p. xiii).

This knowledge gap is part of the World Bank's rationale for the Ganges Strategic Basin Assessment. The data from this study is intended to be the basis of evidence-based policymaking in the near future, and the process of the study itself was an attempt at capacity building and increasing technical and scientific cooperation. It is also the basis of water governance for the South Asia Water Imitative (SAWI).

What is water governance? The Global Water Partnership defines water governance as 'the range of political, social, economic and administrative systems that are in place to develop and manage water resources, and the delivery of water services, at different levels of society' (Rogers & Hall 2003, p. 7).

We now understand good governance, and water governance. What, then is good water governance? Feldman (2007, p. 5) has adapted from Gleick seven criteria of good water governance that aim for the sustainable development of water resources: maintaining human health; maintaining ecosystem health; minimum standards of quality; long-term freshwater renewability; data collection and accessibility; institutional mechanisms for resolving conflict; and democratic decision-making. For UN Water (2014), 'good water governance enables people to participate in decisions affecting their lives. To ensure this, accountability mechanisms need to be in place'.

Good water governance is also, according to Falkenmark et al. (2007), 'absolutely the top priority' for human security, because only good water governance can enhance access to adequate quantity and quality of water for people's well-being. For this reason Jägerskog et al. (2014, p. xi) remind us that '[m]any studies at international level have indeed identified the water crisis not as one of scarcity but as one of governance'. There is a nexus between water governance and water-related security crises – and these are increasingly prevalent.

The Ganges–Brahmaputra problemshed is not generally characterised by good water governance. As Asthana and Shukla (2014, p. 271) argue, '[c]urrent water resource development in South Asia is not sustainable. There needs to be a shift in emphasis on demand management practices and further incentives for conservation of water resources.' They show that a range of water governance strategies and policies have been suggested by various epistemic communities, NGOs, civil society groups and, of course, international actors like the World Bank. But there is no uniform prescription and the system is a complex one (Asthana and Shukla

2014, p. 287). Complex or not, as the Global Water Partnership makes clear, '[u]sually improving governance means reform' (GWP 2003, p. 2). One avenue for reform in the Ganges–Brahmaputra problemshed is amending institutional arrangements. As the World Bank (2014a, p. 32) points out, 'the Ganges is one of the few large international basins in the world with no permanent institutional mechanism that involves multiple riparians'. Institutionalised collaboration across borders and sectors is, for the major international organisations and Western aid donors, a hallmark of good water governance.

Emergence of non-traditional security threats

The past two decades have seen a general increase of interest in non-traditional security threats and collective action problems around the globe. Over the course of the twentieth century, 'the relationship between the human world and the planet that sustains it has undergone a profound change' according to the 1987 report of the World Commission on the Environment and Development. The report points out that the rate of change is outstripping the ability of scientific disciplines and our current capabilities to assess and advise. This, in turn, 'is frustrating the attempts of political and economic institutions, which evolved in a different, more fragmented world, to adapt and cope' (World Commission on Environment and Development 1987, Part III, Chapter 12, Section III, Paragraph 122).

The understanding of the inter-connectedness of human and environmental systems has been seeping out of ecology and science silos and onto the global stage. Now, argues Matthew (2014, p. x), 'we find ourselves on the eve of the Hydrocene, an era in which the well-being of humankind depends in great measure on the resourcefulness we bring to managing the planet's most abundant resources'.

Environmental issues have been slowly moving into the centre of global politics since the latter half of the twentieth century. Milestones such as the Club of Rome, the Stockholm environmental conference 1972, the Brundtland Report (a.k.a. *Our Common Future*) on sustainable development in 1987 and the Dublin and Rio meetings in 1992 have, according to Jägerskog et al. (2014, pp. vii–viii), marked the incremental progress of environmental issues emerging as a security problem of sorts. But while the environmental security discourse in the 1990s was preoccupied with violent conflict (including state-on-state conflict) over scare resources, it has since then moved on to understanding environmental risks as 'threat multipliers' in conflict-sensitive regions (O'Lear et al. 2013, p. 3).

Conflict is increasingly driven not by external state-based aggressors, but rather by internal or local pressures, and the consequences of poverty or socio-political instability, including vulnerability to environmental threats such as water scarcity (Cosgrove 2003, p. 50). For Wolf et al. (2005, p. 94), '[t]hese changes suggest that tomorrow's water disputes may look very different from today's'. Because the very character of conflict is changing along with emergent non-traditional security threats, Zeitoun and Mirumachi (2008, p. 299) suggest that thinking in terms of conflict (and its binary opposite, cooperation) may be unhelpful: in the new security paradigm conflict and cooperation co-exist. In regard to water disputes, Zeitoun and Mirumachi (2008, p. 299) propose that more insight and understanding will be gained in speaking of water-related *interaction* rather than water conflict or water cooperation.

To some extent, the global agenda is beginning to embrace this nuanced understanding of security and conflict. Nonetheless, as O'Lear et al. point out, there are still differences in how non-traditional security threats are understood in developed versus developing countries. Environmental security is largely viewed in Western countries as a 'threat multiplier' in already conflict-sensitive regions. In contrast, developing countries first and foremost consider security implications with regional neighbours when responding to extreme events (O'Lear et al. 2013, p. 3). This is the case in South Asia, where, as discussed in Chapter 8, there is no security community, and India, Nepal, Bhutan and Bangladesh tend to consider security in realist terms (i.e., state-centric; as Jägerskog et al. (2014, p. xv) point out, 'traditional security perceptions have centred on the state and its survival in a realist or neo-realist fashion'). However, as global security threats become increasingly non-traditional, and new forms of conflict emerge, Werz and Conley (2012, p. 33) warn that 'our security can no longer be guaranteed by military strength or economic clout alone, but only by our ability to compel collective action'.

This rethinking of collective and human security is slowly beginning to happen. Davies et al. (2015, p. 267) show that action on global challenges is increasingly taken forward by coalitions of likeminded parties, incorporating sovereign states, cities, civil society actors, businesses, institutional investors, regional and international organizations, and youth movements. Coalition governance is increasingly geared toward achieving 'collective impact' on large-scale challenges. Similarly, the vehicles for dealing with such challenges are increasingly special-purpose, internationally coordinated and long-term mega-programmes. The South Asia Water Initiative (SAWI) is one example of this sort of special-purpose, long-term mega-programme led by an international coalition of like-minded parties.

The once prevalent fear that the wars of the twenty-first century will be about water has been largely dispelled by scholars, but the United Nations (UN Water 2014, p. 8) warns that water is an important collective action problem. 'As countries develop and populations grow,' one recent UN report says, 'the potential demand for water is projected to increase by 55% by 2050. … Water supply crises have been identified by industry, government, academia and civil society as one of the top three global risks' (UN Water 2014, p. 8). Keskinen et al. (2014, p. 39) concur: 'Water is gaining higher political attention locally, nationally, regionally and globally – as it should be.' They add that the need for cooperation at different levels is being increasingly acknowledged. Hydro-diplomacy, which is a vital element of water cooperation, particularly in transboundary settings, is likewise gaining momentum.

The momentum to promote hydro-diplomacy is also evident in the interventions of international organisations and aid donors in the Ganges–Brahmaputra problemshed, especially SAWI. Largely hydro-diplomacy and other approaches for increasing transboundary water cooperation are a response to the perceived rise of non-traditional security threats.

The spectre of climate change

Perhaps the most ominous (or, at least, most prevalent in the public imagination) of the non-traditional security threats is that of climate change. The 2012 United Nations report, *The Future We Want*, reaffirms that 'climate change is one of the greatest challenges of our time ... We are deeply concerned that all countries, particularly developing countries, are vulnerable to the adverse impacts of climate change' (paragraph 190). Because the effects of climate change are, among other things, threatening food security, poverty eradication and sustainable development, the United Nations (2012, paragraph 190) emphasises that 'adaptation to climate change represents an immediate and urgent global priority'. Although this provides a focal point for policymakers, Groenfeldt (2010, p. 579) argues that 'climate change adds urgency to more and bigger technical fixes'.

Within the conventional water management paradigm:

> The inclination is to go 'back to the future' of command-and-control approaches and re-engineer a solution to climate change: Build more dams to store more water, more pipelines to cross basin boundaries and more pumping to tap ever-deeper sources of groundwater.
>
> (Groenfeldt 2010, p. 579)

At the same time, there is a clear trend within the scientific community in favour of ecological, rather than engineering, solutions because these 'lend themselves to unanticipated synergies, rather than conventional responses which often result in unexpected collateral damage' (Groenfeldt 2010, p. 579).

Climate change is an increasingly important focus for the World Bank too. The growth of climate change-related activities supported by the World Bank has been tremendous, and today climate change is the largest share of the Bank's environment portfolio (in 2012, almost $4.6 billion of lending supported climate change adaptation and more than $7.1 billion was directed toward climate change mitigation) (Evans 2015a, p. 64).

Increasing importance of water security in international development

The emergence of climate change and other non-traditional security threats that require collective action has broken what Jägerskog et al. (2014, pp. vii–viii) call the monopoly of the nation-state and their armed forces on the security concept.

Water is also gaining higher political attention globally because it 'may be one of the absolutely most crucial commodities for safeguarding people's wellbeing'. As Wolf et al. (2005, p. 94) point out, '[a]n unprecedented number of people lack access to a safe, stable supply of water' and this has serious repercussions for human security. As such, the debate on water security has broadened from the classic international relations field to new areas of scholarship and policy, including development; rights perspectives on water; legal aspects; and the water, food and energy nexus (Jägerskog et al. 2014, p. xi). Thus, just as water has become an issue of high politics, it is also increasingly treated as a development challenge – and a challenge that is only likely to rise in significance, argue Jägerskog et al. (2014, p. xii).

The significance of water as a challenge for development lies partly in the nexus between water security and food security. As Uitto and Duda (2002, p. 366) point out:

> About 70% of all water withdrawal globally is due to irrigated agriculture, which only represents 17% of the total area under agriculture but produces 40% of the world's food. It is likely that water availability will in the near future limit food production potential, at least regionally.

Access to food, nutrition and the livelihoods in the agricultural sector are key issues for international development. Asthana and Shukla (2014, pp. 21–22) claim that the issues on the nexus of food and water security, as well as energy security, are a particular problem in South Asia, especially in India. Chellaney (2014, p. 621) explains the intricacies of the food–energy–water nexus: 'Food production is closely intertwined with water and energy, whereas water and energy are closely connected with each other. Water is essential for energy extraction, processing, and production, while energy is vital to treat, distribute, and supply water.' Any disruption in the fragile balance in food, water and energy systems is likely to have profound effects on human security and development.

Yet, the water-related development challenge that most readily gained international momentum has been that of sanitation and public health. The Millennium Development Goals (MDG), which defined the international development agenda between 2000 and 2015, aimed to halve the proportion of the world's population without sustainable access to safe drinking water and basic sanitation by increasing access to improved drinking water sources and sanitation facilities (UNU 2013, p. 7). The focus on drinking water and sanitation took primacy over water governance largely because '[s]trong connections were never developed between water development and management and reducing hunger and household poverty' even though efforts had been made to connect integrated water resource management to the full set of the MDGs (UNU 2013, p. 8).

Nonetheless, the MDGs were instrumental in connecting water to other development challenges, and establishing them firmly on the agenda for global action. Sachs (2015, p. 490) argues that these international objectives are important because they mobilise social and political action, and they create a sort of positive peer pressure between countries to act and compete on outcomes. They also provide a focal point for epistemic communities to come together and recommend the best courses of action, and mobilise stakeholder networks (Sachs 2015, p. 490). Moreover, the MDGs resulted in increased aid pledges and financial commitments, and 'by strengthening the focus on the social dimensions of development, they have led to a shift in the type of aid delivered' (UNU 2013, p. 11).

At the same time, the MDGs arguably neglected issues rising to prominence on the international security and environmental conservation agendas, such as climate change, and 'they do not capture the overarching goal of sustainability' or 'take account of how growth can contribute to development outcomes' (UNU 2013, p. 11). The Sustainable Development Goals (SDG) were established in 2015 as the new development agenda up to 2030 in part to correct these omissions, and in part to better reflect the changing landscape of international development (particularly the fact that 'many emerging economies are no longer eligible for official development assistance (ODA) and have become aid donors themselves, and developed countries have become increasingly dependent on capital from developing countries' (UNU 2013, p. 15)).

The MDGs have also been a catalyst for international collective action on various development issues, especially in relation to the provision of global public goods such as water and sanitation (Evans et al. 2015, pp. 189–190). This has changed the way that international institutions, including the World Bank, interact and partner with local non-governmental organisations, non-state and sub-state actors. Together, the World Bank and a large section of the international donor community are increasingly pursuing an agenda addressing global threats, collective action problems and providing global public goods, as well as focusing on regions – such as South Asia – particularly vulnerable to these threats (Davies & Evans 2015, p. 274). This is likely to continue in the post-2015 era of SDGs.

One of the SDGs is dedicated to water management, including cooperation over transboundary water resources, and stresses equitable distribution (UN 2015a, Goal 6). Because water supply crises 'have been identified by industry, government, academia and civil society as one of the top three global risks', the inclusion of water management as one of the SDGs is likely to 'send a clear message that policymakers should focus on this impending threat' (UN Water 2014, p. 8). This has indeed been the case, and the international attention on transboundary water governance in the Ganges–Brahmaputra problemshed is one manifestation of this growing policy focus on water related risks. The importance of the SDGs, therefore, is not that they include a water-focused goal, or that they somehow directly drive foreign-led interventions in the Ganges–Brahmaputra problemshed. Rather, the significance lies in the SDGs being one prominent manifestation of the burgeoning international interest in water, water security, water governance and the sustainable development of water resources. The discourse on water as a challenge for international development informs international actors, and the emerging global consensus on the principles of addressing these challenges drives them – in conjunction with other drivers discussed in this chapter – to intervene in transboundary water in the Ganges–Brahmaputra problemshed.

Sustainable development and poverty eradication

As the name suggest, the Sustainable Development Goals are also significant in cementing the concept of sustainable development on the international agenda. Sachs (2015, p. 1) sees sustainable development as 'both a way of understanding the world and a method for solving global problems'. It is an idea that is quite familiar to many people the world over now, but this has been a long time coming. Although, as Allan (2006, p. 39) points out, '[t]he water sector in the North adopted principles of sustainable development during a protracted discursive struggle from the late 1950s to the mid-1970s', the concept has not gained widespread acceptance until recently.

By 2012, the Rio+20 conference accepted that 'water is at the core of sustainable development as it is closely linked to a number of key global challenges. We therefore reiterate the importance of integrating water in sustainable development' (UN 2012, paragraph 119). Water for sustainable development has been defined as 'the use of water that supports the ability of human society to endure and flourish into the indefinite future without undermining the integrity of the hydrological cycle or the ecological systems that depend on it' (Gleick cited in Mays 2007, pp. 4–5). As such, a sustainable water resources system is one 'designed and managed to fully contribute to the objectives of society, now and in the future, while maintaining their ecological, environmental, and hydrological integrity' (Loucks 2000, pp. 3–10). The importance of the integrated approach to managing water for sustainable development is echoed in the SDGs, especially Goal 6.5: 'By 2030, implement integrated water resources management at all levels, including through transboundary cooperation as appropriate' (UN 2015b, p. 18). The explicit mention of transboundary cooperation is also significant. It underscores the rising importance of transboundary water governance on the international agenda, and is another push factor for international organisations and foreign aid donors in the Ganges–Brahmaputra problemshed.

Tied to the idea of sustained and sustainable economic growth in the discourse of sustainable development is the idea of poverty eradication. Indeed, the report *The Future We Want* states '[e]radicating poverty is the greatest global challenge facing the world today and an indispensable requirement for sustainable development' (UN 2012, paragraph 2). For Sachs (2015, p. 3), by definition 'sustainable development calls for a world in which ... extreme poverty is eliminated'. Moreover, poverty eradication should, according to several United Nations institutions, be one of the guiding principles of the SDGs – along with economic growth, equity and, of course, sustainability (UNU 2013, p. iv).

The objective of poverty eradication puts the onus of responsibility for sustainable development on rich countries and international institutions. Back in 1987, the *Our Common Future* report called on international organisations, particularly the World Bank, to take a leadership role on sustainable development. Developing countries need a significant increase in financial support from international sources for environmental restoration, protection and improvement, the report argued. This was seen as necessary to help these countries through the necessary transition to sustainable development. Moreover, international organisations were seen as key to supporting this transition. The authors of *Our Common Future* found it difficult to envisage developing countries making this transition to sustainable development in an effective and timely manner without the full commitment and support of international organisations and agencies (World Commission on Environment and Development 1987, Part III, Chapter 12, Section II, Paragraphs 6.2.100–102).

Good governance, and good governance of water resources, is an integral component of sustainable development, as well as a significant guiding principle of international actors in the Ganges–Brahmaputra problemshed. This was discussed at the beginning of this section. What is interesting about the Sustainable Development Goals is that in emphasising the good governance of water resources, they also make the explicit link (in Goal 6.5) with transboundary water cooperation (UN 2015b, p. 18).

The importance of transboundary cooperation and good governance to sustainable development cannot be underestimated, but as Plummer (2005, pp. 34, 37) points out, '[i]mplementation strategies for sustainable development generally focus on the value of a "top-down" or "bottom-up" approach', yet 'sustainable development requires an approach that is simultaneously top-down and bottom-up'. Plummer (2005, p. 38) asserts that just as a top-down approach is not the sole path to sustainable development, local actions alone are not a panacea either. Sustainable development is, in other words, a collective action problem that requires – like the non-traditional security threats described above – innovative, collaborative approaches: 'for sustainability requires a new governance mode that builds on knowledge of social, ecological, economic, and technological complexity, while understanding how such uncertain processes can be triggered and steered toward cooperative and sustainable provision of the GPGs [Global Public Goods]' (Loorbach et al. 2015, p. 234). In the perceived lack of direction on these collective action problems in the Ganges–Brahmaputra problemshed, international actors are stepping in to take the lead.

'Command and control' vs. ecological principles in environmental resource management

The emphasis on sustainable development reflects a changing understanding of and approach to the environment. That nature could and should be controlled was, according to Allan (2006, p. 40), the dominant paradigm for 'both capitalist and socialist versions of industrial modernity during the first 75 years of the 20th century'. Groenfeldt (2010, p. 576) shows that in the nineteenth and early twentieth centuries water management in particular was based on the 'command and control principle'. Dams were built to impound rivers and release the water for power generation, irrigation, industry or urban water supply. Canals were built to move water where rivers did not flow. At the same time rivers were straightened and deepened for ease of river transport and to evacuate flood waters faster, while levees kept rivers contained and away from their natural floodplains.

State agencies and entrepreneurs involved in delivering water for economic and social purposes believed that nature in general, and water specifically, could 'be subject to the mastery of science and industry' and this, writes Allan (2006, p. 42), 'was possible because of the revolutions in science and industry in the early 19th century and the achievements of capitalist organisation in marshalling the resources of labour, the environment and capital'.

Significantly, the command and control principle was and still is, Groenfeldt (2010, p. 576) argues:

> intricately intertwined with the principle of 'beneficial use' and, most particularly, economically beneficial use. Under the conventional paradigm, water should be tightly controlled in order to direct its service to a 'stream' of economic benefits. If water is not providing economic benefits for somebody, it is not being put to beneficial use.

The pursuit of maximised economic benefits has led, claim Kibaroglu et al. (2008, p. 226), to an environmental crisis that is also the second failure of capitalism (the first being the mismanagement of labour) because it is a 'result of the reckless use and abuse of too many resources by too many people'. In recent years, however, the values and principles guiding environmental resource management have been shifting; since the 1980s and 1990s, the discourse has become, according to Groenfeldt (2010, p. 576), more complicated and nuanced, and in the past decade in particular the role of nature has taken on a different tone.

The rights of nature are now increasingly considered in addition to human rights and human development (Groenfeldt 2010, pp. 575–576). Nevertheless, this paradigm reflecting environmental concern has, according to Allan (2006, p. 41), only achieved very limited purchase on water policymaking in the South. As discussed in Chapter 2, the command and control approach to water resources management is still prevalent in the Ganges–Brahmaputra problemshed.

There is, Groenfeldt (2010, pp. 577–578) argues, far more rhetoric than action in adopting ecological principles. Water related crises, he adds, continue to be defined as a series of physical challenges to be solved through engineering, even though the conventions of water management are slowly shifting towards an emphasis on ecosystem services and sustainability. The concept of sustainability has certainly been useful, claims Allan, in ensuring the reliable delivery of environmental services and raising the profile of the importance of the environment, but it has proved to be impossible to operationalise in a narrow environmental sense. This is because the environment is being conceptualised as a separate focus of policy rather than one integral to social and economic uses of water in the political economy as a whole (Allan 2006, p. 49). In other words, there has not yet been a broad and committed adoption of what Groenfeldt (2010, pp. 577–578) calls ecological water management, which 'builds on natural riparian principles and ecological relationships that are becoming increasingly well understood' and 'offers a way of working with, rather than against, nature's water cycles of precipitation, flow, and infiltration'.

One significant policy application of ecological water management could be a commitment to environmental flow. A minimum environmental flow is the smallest amount of water required at any given time to allow the proper functioning of the natural water regime of a river, wetland or coastal zone which maintains the ecosystem. Groenfeldt (2010, pp. 577–578) argues that '[e]nvironmental flows provide critical contributions to both river health and ultimately to economic development, ensuring the continued availability of the many benefits that healthy river and groundwater systems bring to society'. This sort of approach, however, has not been prevalent in the strategies of international actors engaging in transboundary water governance in the Ganges–Brahmaputra problemshed, despite the rhetoric around sustainable development.

6 From water conflict to water cooperation through the World Bank?

'Fine minds in South Asia are seeking solutions for the looming water and river crisis in this region,' argues Colopy (2012, p. 319). She points out that:

> Some outsiders would like to help. Many feel genuine concern about the rivers and people of South Asia ... Perhaps South Asia could use help from outsiders, perhaps not; but governments in the rest of the world are starting to pay attention.
>
> (Colopy 2012, p. 319)

Are these outsiders likely to improve transboundary water interaction in the Ganges–Brahmaputra problemshed? That is the central question of this chapter.

International actors are, as discussed thus far, driven by different factors to intervene in the transboundary water governance of the Ganges–Brahmaputra problemshed. There are the numerous water related crises in the region that the riparian states are seemingly unwilling or unable to address. These were examined in Chapters 2 and 3. There are also external drivers of foreign-led interventions, such as geopolitical factors and dominant international paradigms (for example, the concepts of good governance and sustainable development). These were discussed in Chapters 4 and 5.

International organisations and foreign aid donors are also interested in increasing transboundary water cooperation as a way of decreasing water conflicts. The Ganges–Brahmaputra problemshed is particularly vulnerable to water conflicts, which are generally understood to be the result of state fragility (recent violence, obstacles to economic development, weak management institutions), and high water stress (National Research Council of the National Academies 2012, p. 92). As water governance issues gain higher political attention locally, nationally, regionally and globally, cooperation and hydro-diplomacy are increasingly seen as means of attenuating water-related disputes, particularly in transboundary settings (Keskinen et al. 2014, p. 39). This is one of the explicit objectives of the South Asia Water Initiative (SAWI) and the governments of Australia, United Kingdom and Norway in their facilitation of Track II dialogue in the Ganges–Brahmaputra problemshed. This chapter examines this and other explicit objectives of international actors in the Ganges–Brahmaputra problemshed.

First, however, this chapter establishes what the main international interventions are, and how they came about.

In 2006, the Abu Dhabi Dialogue (ADD) was established for increasing water cooperation in the Hindu Kush Himalayan region. SAWI eventually arose out of this process. The increasing activities of the World Bank in regard to water governance in South Asia paved the way for several international interventions. The genesis of these interventions, and how they work toward improving transboundary is discussed in the first section of this chapter.

The convergence of the discourses discussed in Chapter 5 is evident in the explicit objectives of these foreign-led engagements in transboundary water governance in the Ganges–Brahmaputra problemshed. These objectives are four-fold: establishing good water governance; ensuring overall regional stability; supporting climate change resilience; and maintaining sustainable development and poverty eradication. Knowledge sharing is also an important aim that underpins all the others. These interlinked objectives are discussed in turn.

The ADD and SAWI

'Considering the enormity of the climate change challenges in South Asia,' claim Earle et al. (2015, p. 41), 'there is at present no paucity of collaborative efforts to explore effective response strategies'. In the mid-2000s, the confluence and culmination of these climate change challenges with the factors described throughout Chapters 2 and 3 resulted in an international round of talks called the Abu Dhabi Dialogue (ADD). The ADD periodically brought together senior members of government, academia and civil society from Afghanistan, Bangladesh, Bhutan, China, India, Nepal and Pakistan in a non-formal, consultative process intended to foster regional cooperation on various challenges facing Himalayan Rivers (SAWI 2015c). Experts from the region also met at various international conferences (Earle et al. 2015, p. 41). An issue of specific concern was 'climate change and its future impacts on livelihoods from the headwaters to the floodplains' (SAWI 2015c). The ADD also funded small grant projects for non-government organisations (NGOs) (Hill 2013, pp. 255–256).

The vision of the ADD was medium term, with agendas looking a decade in advance (Vaidya & Sharma 2014, p. 4). The ADD was intended to be a neutral forum for open conversations among participants. Statements made in the forum were non-attributable and there was no requirement for a consensus outcome (SAWI 2015c). Earle et al. (2015, p. 41) describe the ADD as the only multilateral framework involving all the riparian countries of the Ganges–Brahmaputra basin that was engaged in informal and consultative processes. They point out that it tried to build trust between riparians primarily through information exchange and the co-management of development projects (Earle et al. 2015, p. 41). This is an important undertaking because transboundary water management is a politically sensitive topic in South Asia, yet one that is not necessarily a top political priority for all riparians. Moreover, there is little trust between these riparians.

The ADD is of significance here because it is the genesis of concerted efforts by international actors to improve transboundary water governance in the Ganges–Brahmaputra problemshed and South Asia more broadly. The first round of talks (called the 'First International Conference on Southern Asia Water Cooperation' and held in Abu Dhabi in September 2006) was hosted by the International Institute of Strategic Studies (IISS) (Earle et al. 2015, p. 41) with the support of the UK Foreign and Commonwealth Office (SAWI 2015e). At these talks, participants of the Abu Dhabi Dialogue Group (that is, representatives from Afghanistan, Bangladesh, Bhutan, China, India, Nepal and Pakistan) decided that the World Bank should facilitate sustained dialogue on rivers of the Greater Himalaya region (SAWI 2015e). For the World Bank, this was an opportunity to create a 'common water-sharing framework in line with the Nile Basin Initiative' (Earle et al. 2015, p. 41). The Nile Basin Initiative (NBI) has not been an unmitigated success in terms of increasing transboundary water cooperation (Swain 2011, p. 696) and so perhaps there is an element of the World Bank wanting to apply the lessons learned from the Niles basin to another part of the world (the Ganges–Brahmaputra is an appropriate choice as there is no other basin-wide collaborative process but the potential for conflict remains high).

Since its inception, the Abu Dhabi Dialogue Group (ADDG) has convened six times (SAWI 2015c). In 2007, the seven countries met to 'share global experience on international waters and benefit-sharing, with the goal of achieving constructive convergence on common challenges' (SAWI 2015d). The language of benefit sharing and common challenges to be overcome collectively reflects that of the dominant international narratives discussed in Chapter 5. The 2007 meeting of the Abu Dhabi Dialogue Group is also when the South Asia Water Initiative (SAWI) was established by the World Bank to 'support future ADDG activities' and to 'promote the goals of poverty reduction, economic growth, mitigation and adaptation to climate change, water security and regional peace through significant and measurable improvements in water resources management and development' (SAWI 2015d).

Significantly, SAWI is funded exclusively by external donors (Australia, United Kingdom, Norway) (World Bank 2014b, pp. 30–34) without even symbolic contributions from the Abu Dhabi Dialogue Group member states. The Nile Basin Initiative also did not receive funds from riparian states, and Swain points out that its almost total dependence on external funding for its inception and operation undermines its long-term sustainability. In Swain's (2011, p. 696) opinion, this dependence on external funding reflects a lack of commitment from the basin countries to basin-based water management. How this affects the efficacy of SAWI is discussed below.

In 2008, country level dialogues were held in India, Nepal and Bangladesh and the ADD formed the Abu Dhabi Knowledge Forum under the auspices of the International Centre for Integrated Mountain Development (ICIMOD) (SAWI 2015f). The Knowledge Forum convened more than 50 institutions from across the region and was thus more inclusive than the Abu Dhabi Dialogue Group, but the two are complementary. The main function of the Knowledge Forum is knowledge creation and knowledge sharing. Again, the focus on evidence-led decision-making and data driven policy echoes the prevailing discourses discussed in Chapter 5. This theme was reiterated in the 2009 ADD during which 'it was agreed that greater data sharing could help to catalyze further regional cooperation' (SAWI 2015g). To this end, the Small Grant Program was established in 2011 to 'initiate new knowledge generation, expand current national research activities to extend across boundaries, and disseminate knowledge within the region' (SAWI 2015j).

In addition to the focus on knowledge creation and sharing, there are three other recurring themes of the ADD. These are: lessons learned from other international basins (e.g., the Mekong), the application of international water law to the transboundary basins of South Asia (SAWI 2015h), and the deliberative/consultative processes required for collective action (SAWI 2015i). These are all recurring themes throughout the work of SAWI and the other foreign-led interventions in the transboundary water interaction of the Ganges–Brahmaputra problemshed.

At the fourth Abu Dhabi Dialogue (held in 2009), an international, interdisciplinary team led by World Bank economists embarked on a 'multi-sectoral study to understand the Ganges basin, its present behaviour and possible futures' (Sadoff 2009). The study synthesises publicly available data (while acknowledging the limitations of the availability of such data) into 'a set of nested hydrological and economic river basin models used to examine alternative scenarios across a range of possible Ganges futures, and a social component that studies the social implications of water variability in the basin' (World Bank 2014a, pp. xiii, 2). The Ganges Strategic Basin Assessment (GSBA) manifests the international values of data-led policy and evidence based decision-making, and as such provides the basis for what could be the main thrust of western-led solutions to transboundary water governance in the Ganges basin.

The Ganges Strategic Basin Assessment, and ADD process have been informed and influenced by the paradigms and discourses discussed in Chapter 5. They are recurring themes in the strategic planning documents and programme rationale documents of these interventions. The explicit objectives of many of these international organisations and foreign aid donors can be grouped into four main (and overlapping) categories, namely:

1. Instituting good governance and hydro-diplomacy
2. Ensuring overall regional stability
3. Supporting climate change resilience
4. Maintaining sustainable development and poverty eradication.

Underpinning each objective is knowledge sharing and data creation. Each objective is discussed below.

Establishing good governance and hydro-diplomacy

As discussed in Chapter 5, the prevailing international paradigm of governance favours multilateral, deliberative processes that aim for consensus and result in positive-sum outcomes. As UN Water points out, fundamental to the water goal and sub-goals of the Sustainable Development Goals (SDG) is good governance based, among other things, on participatory decision-making, regulatory frameworks, knowledge transfer and skills development. Moreover, 'improved governance can be viewed as a pre-condition for the successful achievement of the ... water targets' (UN Water 2014, p. 26).

To recap, water governance is a social process of dialogue, negotiation and decision-making (Dore 2014, p. 194). Deliberative water governance, then, is the constructive engagement in water governance through the promotion of 'inclusive, deliberative processes that emphasise different perspectives, critical analysis, learning and institution-building whilst respecting rights, accounting for risks, acknowledging responsibilities and fairly distributing rewards' (Dore 2014, p. 195). According to the United Kingdom's Department for International Development (DFID 2012), a deliberative approach is especially relevant in transboundary water governance because 'decisions involve many stakeholders, complex issues, trade-offs between policy options, high levels of uncertainty and low levels of trust' and it can 'help resolve complexity by building trust, promoting a shared understanding amongst stakeholders of the issues (in particular where these issues are common across national boundaries)'. This is the theory of change on which the objectives of international actors intervening in the Ganges–Brahmaputra problemshed is based.

The influence of this paradigm is evident in the stated goals of the South Asia Water Initiative (SAWI), as well as the strategies of the United Kingdom and Australia vis-à-vis transboundary water governance. The main objective of SAWI is to promote regional cooperation in the sustainable management of water resources of the Himalayan rivers based on the principles of Integrated Water Resource Management (IWRM), inclusive and multi-disciplinary dialogue, and participatory and deliberative processes (SAWI 2015a). For Australia – a key partner in SAWI – the goal is to promote transboundary water resource management in line with the Department of Foreign Affairs and Trade's (DFAT) South Asia strategy (SAWI 2015b). The United Kingdom, another key partner in SAWI, also aims for increased regional cooperation in the management of the Himalayan rivers through the establishment of transboundary and national governance arrangements that move away from a technical approach to a deliberative approach (DFID 2012).

A further objective is 'to help achieve the mutual benefits of transboundary cooperation across shared river basins' (SAWI 2015b). A key principle of SAWI is to deliver mutual benefits (SAWI 2015a). The theory of change here is that mutual benefits accrue:

> from coordinated planning and joint investment for new infrastructure that ensure equitable upstream–downstream sharing of both benefits and impacts,

> and from coordinated planning and joint investment into upstream watershed rehabilitation that deliver both local benefits and downstream water quality benefits.
>
> (SAWI 2015b)

Track II dialogue, which was discussed Chapter 1, is a mechanism of deliberative governance favoured by SAWI and its donor states in the Ganges–Brahmaputra problemshed. Deliberative processes such as Track II dialogue 'are a way of involving people in decision-making' according to DFID (2012), and '[a] deliberative approach is particularly relevant to the question of trans-boundary water governance, as decisions involve many stakeholders, complex issues, trade-offs between policy options, high levels of uncertainty and low levels of trust'. This is also the basis of hydro-diplomacy.

Ensuring overall regional stability

Related to hydro-diplomacy is the goal of regional stability. As discussed in Chapter 5, the emergence of non-traditional security threats and collective action problems such as climate change has shifted the way in which regional security is addressed. Conflict, be it over water or otherwise, is increasingly being driven by internal or local pressures, or subtler issues of poverty and stability, and thus becomes less traditional (Cosgrove 2003, p. 50). Moreover, these non-traditional security threats do not respect international boundaries, and sustainable security requires collective action that addresses human security and rebalances the three tools of foreign policy – defence, diplomacy and development (Werz & Conley 2012, p. 34).

In the Ganges–Brahmaputra problemshed, low transboundary water cooperation poses a threat to regional stability because poor water management imposes economic and social costs, leads to inefficiencies in water supply and increases vulnerability to floods and droughts, which in turn can create disputes that, if left unresolved, play into broader political tensions. Water issues can therefore exacerbate existing disputes, such as those found between Bangladesh and India on the Ganges basin, and in turn may increase the potential for escalating tension both within and between countries (DFID 2012). Water related disputes can and do occur within as well as between countries.

The theory of change for international actors intervening in this problemshed is that by increasing cooperation between riparian states there will be a net stabilising effect and that in turn will reduce the likelihood of regional conflict, especially conflict related to water. The World Bank has a political agenda for creating general regional stability in South Asia (Molden 2014). The overarching objective of SAWI (2015a) is therefore to increase regional cooperation (and the goals of facilitating deliberative water governance and increasing climate change resilience support this). Regional stability is also an explicit goal of Australia's climate change adaptation programmes (AusAID 2013, p. 4).

This is a theory of change that the United Kingdom also subscribes to. The idea is that better transboundary river management will open up communication channels between countries and create greater interdependence between them, thus leading to greater regional stability (Anonymous 1, 2014). Indeed, the conflict pool strategy of the United Kingdom's Foreign and Commonwealth Office, the Ministry of Defence and DFID, 'considers transboundary waters as a key issue for reducing security risks in the region' (SAWI 2015b). The DFID South Asia Water Governance Programme (SAWGP) works in support of the United Kingdom's Conflict, Humanitarian and Security Department (CHASE), including by helping in the identification of emerging risks and in pre-empting crises in a strategically important region (DFID 2012). The strategic business case for this programme claims that 'improved cooperation on water and energy will reduce the potential for water and energy issues to act as irritants to bilateral relations, reducing tensions and enhancing security across the region', and expects regional stability to be a co-benefit in addition to improved water governance and the building of climate resilience (DFID 2012).

Supporting climate change resilience

The third explicit objective of international actors intervening in the transboundary water governance of the Ganges–Brahmaputra problemshed is to build climate change resilience. As discussed in Chapter 2, South Asia is particularly vulnerable to the effects of climate change. Climate change is already changing precipitation patterns, increasing floods and droughts, causing Himalayan glaciers to melt at unprecedented rates, and thus adversely affecting tens of millions of people in this poor, densely populated and predominantly agrarian region.

The unique challenge for the Ganges–Brahmaputra–Meghna mega-basin is that monsoon rain makes river flows highly variable, and climate change will further exacerbate this variability. At the same time, rapid population growth and unregulated urbanisation will increase demand for water from less predictable rivers – a governance challenge that may be particularly devastating considering that the region's inability to manage current climate variability bodes poorly for tackling the climate change impacts of the future (DFID 2012). Indeed, the Ganges–Brahmaputra problemshed is ill-prepared institutionally for climate change. For example, none of the treaties and agreements that Nepal has with India address climate change or the uncertainty posed by potential effects of changing melt dynamics from Himalayan glaciers (National Research Council of the National Academies 2012, p. 86).

The World Bank has a political agenda that includes climate change adaptation (World Bank Group 2008) and that is reflected in the objectives of the World Bank-led SAWI, which addresses the challenges of development and climate change. One of the guiding principles of SAWI is an integrated water management approach that encompasses adaptation to climate change (SAWI 2015a). Similarly, the Australian objective in supporting SAWI is 'to increase climate resilience' and its funding originally came from the AusAID International

Climate Change Adaptation Initiative (Rollason 2012). Indeed, the overall Australian strategy for the development of South Asia makes an explicit link between climate change and adverse consequences for social, political and economic stability (AusAID 2013, p. 4). Likewise, the United Kingdom's South Asia Water Governance Programme (SAWGP) is seeking to achieve greater climate resilience in the region (DFID 2012). For Norway, promoting low-carbon development, with particular emphasis on renewable energy and sustainable management of natural resources, is part of a broader international campaign for 'greener development' (Government of Norway 2011, p. 2).

Maintaining sustainable development and poverty eradication

Greener development, or sustainable development, is a top priority for international organisations and aid donors on the post-2015 international development agenda. The discourse of sustainable development has also shaped the logic of international actors in the Ganges–Brahmaputra problemshed.

The intimate relationship between sustainable environmental management and human development is of particular importance to the Norwegian justification for involvement in the transboundary water governance of this problemshed. Indeed, Norway has been an international champion for sustainable development since former Norwegian Prime Minister Gro Harlem Brundtland chaired the United Nations' World Commission on Environment and Development in 1987.

The Government of Norway's 2011 white paper on Greening Development is unequivocal on the inextricable links between environment and development, and what should be done about it: 'Lasting progress in the fight against poverty depends on long-term economic growth and equitable distribution of social and economic goods. Without radical restructuring of the world economy, continued global growth will cause serious environmental damage' (Government of Norway 2011, p. 2). This commitment to sustainable development is also part of the rationale for Norway's contribution to SAWI (2015b).

For the Australian and UK governments (DFID 2012), the decision to engage in transboundary water governance in the Ganges–Brahmaputra problemshed is also based on the understanding that sustainable environmental resource management is part of the solution to development challenges. Australia's regional development strategy for South Asia is based on two pillars, the first of which is sustainable development. The number one priority under this pillar is to 'support the region in promoting greater water security through collaborative trans-boundary water resource management, particularly through transferring Australia's best-practice technology and science, and strengthening policy and civil society discussion' (AusAID 2013, p. 11). Australia is addressing sustainable development challenges in the region through the Sustainable Development Investment Portfolio, which targets a sub-region of South Asia defined by three major Himalayan river basins – the Indus, Ganges and Brahmaputra – and which aims to increase water, food and energy security in the region to facilitate economic growth and improve livelihoods (DFAT 2014, p. 6). Poverty alleviation in South Asia is also a major concern for the Australian Government (AusAID 2013, p. 8).

For SAWI, the delivery of 'sustainable, fair and inclusive development' is the overarching objective (SAWI 2015a). For the United Kingdom in particular, sustainable development and improved water governance are linked to poverty reduction (DFID 2012). The DFID South Asia Water Governance Programme, for example, is explicitly based on the theory of change that its support for 'more sustainable and more accountable policy decisions and investments' in the water governance sector will reduce overall poverty in the region (DFID 2012). These poverty eradication and sustainable development objectives are in all cases linked to the goal of climate change resilience.

Knowledge sharing

In addition to promoting these objectives and facilitating Track II dialogue, one of the main tasks of third parties in the Ganges–Brahmaputra problemshed is the gathering and dissemination of hydrological data. As discussed in Chapter 7, there is not much accurate data about the availability of resources, their distribution over space and time, and the impacts of climate change on stream flow variability, sedimentation, potential GLOF events and so on (Vaidya & Sharma 2014, p. vi). What information does exist is not readily accessible to those who might need it. Prasai and Surie (2015, p. 28) document this in their 2015 report *Water and Climate Data in the Ganges Basin: Assessing Access to Information Regimes and Implications for Cooperation on Transboundary Rivers*.

The dearth of data has been taken up by international actors as an opportunity for intervention in the transboundary water interaction of the Ganges–Brahmaputra problemshed. Asthana and Shukla (2014, p. 274) see some merit in this approach. They argue that '[a] regional information base where all countries can have access to information and take adequate steps to mitigate damages can be made to remove problems of data secrecy'. Making '[s]ignificant advances in the generation and use of relevant data, knowledge, and analysis' about the rivers of the Hindu Kush Himalayan region is one of the strategic goals of the ICIMOD (ICIMOD 2016), and many of the successes that SAWI claims for the 2013–14 financial year relate to river basin modelling projects (World Bank 2014b). The bulk of the international aid money is funnelled into politically blind areas, that is, science, technology and river modelling (Prasai 2014). This is justified on the grounds that accurate data is critical for planning and monitoring purposes. Scientific data is seen a critical aspect of transboundary water cooperation because without appropriate data exchange, there cannot be collaboration (Molden 2014).

Data creation is an appealing type of intervention for international organisations and foreign aid donors because it is relatively straightforward to design data gathering projects and programmes. Data is quantifiable and therefore progress can be easily measured (e.g., in terms of basins modelled, or flow data points recorded). It is also easy to claim that knowledge creation is a politically neutral endeavour, one for the benefit of all. This is the basis of much of the work done by the ICIMOD, which is able to bring together scientists and researchers from the region.

7 Mixed reactions to foreign-led initiatives from riparian stakeholders

Understanding how policymakers, water security analysts, hydrocrats and water justice activists within the Ganges–Brahmaputra problemshed perceive foreign-led interventions is important in assessing how likely they are to achieve their goal of improving transboundary water interaction. These stakeholders have mixed opinions regarding the utility and efficacy of foreign-led interventions into transboundary water governance, especially the South Asia Water Initiative (SAWI). These mixed opinions are the subject of this chapter.

There is resentment within the problemshed about the top-down approaches of the World Bank and other aid donors, but some reluctant acceptance that there may be advantages to the involvement of a neutral third party. At the same time, however, India has not been an enthusiastic participant in the Abu Dhabi Dialogue (ADD) and SAWI, and this undermines the credibility and effectiveness of these foreign-led interventions. Last, this chapter considers what the rejection by riparians of the initial Ganges Strategic Basin Assessment (GSBA) suggests about the success of third parties in leading Track II dialogue in this region.

Resentment about top-down approaches

The Abu Dhabi Dialogue and subsequent interventions of international organisations and foreign aid donors into transboundary water governance have come at a time of growing nationalism among the countries of South Asia. This nationalism is marked by resentment of that which implies putting aside perceived sovereign rights in favour of multilateralism. As Pachova et al. (2008, p. 289) argue, transboundary water cooperation has been traditionally seen and interpreted as infringing the sovereignty of riparian states.

In the Ganges–Brahmaputra problemshed too there is resentment and distrust of that which may impinge (or be perceived to impinge) on sovereignty and independence. Likewise, in this previously colonised region there is a widespread distrust of any initiative that could even remotely be construed as neo-colonial. This includes foreign-led intervention in transboundary water governance. An instance of this can be found in the reaction from Nepal, Bangladesh and even India toward the World Bank's Ganges Strategic Basin Assessment. This comprehensive report was perceived by these countries as an external imposition on their sovereign right to manage their own water resources (Pun 2013, p. 12). The Ganges Strategic Basin Assessment, and by extension SAWI, are both seen as failures because they are both top-down initiatives (Anonymous 9, 2014). The attempt to impose a bottom-up, deliberative process from an external source in a top-down way is sometimes perceived as neo-colonialist and thus resented within South Asia (Anonymous 11, 2014).

There are also objections from within the Ganges–Brahmaputra problemshed to the top-down drive for what is in essence a bottom-up process – deliberative and inclusive governance. For example, the International Union for the Conservation of Nature (IUCN) Bangladesh–India Initiative is considered by some analysts to be more successful than SAWI at initiating government-to-government cooperation because it is a bottom-up process and thus embodyies the principles of deliberative governance (Anonymous 9, 2014). There is more buy-in from stakeholders for the IUCN initiative and this may be because, though funded by international donors, it is essentially a local solution. Facilitating local solutions and letting host countries develop on their own has, according to Riddell (2014, p. ii), been shown to be a more effective approach to international development than the imposition of solutions from the outside, which tends to create resentment towards the aid donors and undermine their work.

Instead of facilitating local solutions, SAWI is interested in bringing to the Ganges–Brahmaputra problemshed solutions driven by global discourses on good governance, climate change resilience, IWRM and so on (this was discussed in Chapter 5). As such, one analyst (Anonymous 10, 2014) has observed that 'when all you have is a hammer, then every problem becomes a nail'. The implication is that a basin-wide approach, Track II dialogue and evidence-led policymaking – all buzzwords of foreign aid donors and international organisations – may not be always appropriate, or appreciated, in the context of the Ganges–Brahmaputra problemshed.

The resentment about the top-down initiatives is compounded by resentment toward the perceived imposition of 'foreign' ideas, especially through the system of international development aid, which is widely perceived as flawed, corrupt and inefficient (Riddell 2014, p. ii). For a start, the language of the international development sector is seen by some analysts in the region as unrealistic. It tends to focus on outputs and outcomes, rather than the idea that progress is a long term struggle and can be nebulous and unquantifiable (Prasai 2014). Yet, the work of SAWI, Australia, the United Kingdom and so on in the Ganges–Brahmaputra problemshed is based on the same assumptions that most international development projects are based on: that quantifiable progress can be made within one funding cycle, and that there will be outputs and outcomes directly proportional to the amount of money expended (Prasai 2014).

Transboundary water governance in the twenty-first century is moving away from unilateral action to shared strategies for multilateral cooperation and putting human development at the centre of water governance. This is also increasingly true for South Asia, although conventional approaches to water governance (which have prioritised the role of the state in decision-making and management of water resources and been dominated by technical experts and hydrocrats rather than civil society engagement) are still prevalent (Prasai & Surie 2013, p. 5). Moreover, some within the Ganges–Brahmaputra problemshed see the ideal of cooperation through dialogue a patronising and naïve one (Anonymous 4, 2014).

A further criticism of the international interventions in the transboundary water governance of this region is that, according to one western bureaucrat (Anonymous 10, 2014), they are 'a lot of solutions looking for problems' while at the same time these 'solutions' are predicated on the flawed assumption that quantifiable progress towards a lofty goal can be made within one funding cycle (Prasai 2014). By others, SAWI is perceived as ineffective because it focuses almost exclusively on facilitating meetings rather than on projects with tangible outcomes (Anonymous 7, 2014).

Critics of the way in which deliberative governance is being implemented in the Ganges–Brahmaputra problemshed also come from within the ranks of international organisations and proponents of liberalism, that is, those who see international institutions as the most appropriate agents of global governance (Dinar 2002, pp. 241–242). Many are well aware that the facilitation of a broad, deliberative process is still not a core competency for the World Bank (Anonymous 1, 2014), despite the shift in recent years (discussed in Chapter 5) away from funding major infrastructure projects and toward capacity building, good governance and so on (Evans et al. 2015, pp. 190–191). Colopy (2012, pp. 162–166) argues that in some cases World Bank funding is not only not resolving water crises, but it is actually adding to the problem.

Reluctant acceptance that there may be advantages to some foreign-led interventions

Despite these shortcoming and criticisms, the World Bank and other donor agencies intervening in the transboundary water governance of the Ganges–Brahmaputra problemshed are also seen to be doing that which no other organisation is doing. For example, SAWI is actively trying to involve China in the dialogue about the management of the Brahmaputra and that is important and laudable (Thakkar 2014). (Despite the anti-China sentiment in Indian media, and the prevalence of commentators such as Brahma Chellaney who warn of an impending water-war instigated by China, Hill (2015, p. 735) argues that there are some signs of increasing water cooperation. He writes that while China is not moving as quickly towards forging cooperative mechanisms as might be desirable, there are indicators of encouraging shifts towards an emerging broader regional engagement. These include a treaty with lower riparian Kazakhstan, a longstanding role as a dialogue partner in the Mekong River Commission, and some encouraging agreements signed for the sharing of data with India.

There is also a significant need and appreciation for the knowledge that SAWI and the other donor-led interventions (particularly the more scientific ones such as the International Centre for Integrated Mountain Development (ICIMOD)) are producing, especially as that information remains in the public domain (Anonymous 5, 2014). Capacity building is another valued contribution of the World Bank to the region. As the World Bank's role decreases in funding infrastructure projects, it is shifting its focus onto capacity building in water resource management and that is a very welcome change for some hydrologists and hydrocrats in the Ganges–Brahmaputra problemshed (Ray 2014).

A further perceived benefit of international actors in the water governance of the Ganges–Brahmaputra problemshed is that they are often seen as a neutral third party. They have a role to play in facilitating dialogue or mediating between actors who may otherwise be distrustful of each other (Mittra 2014). For all its faults, the World Bank is also sometimes perceived as providing the 'adult supervision' to the 'children' of South Asia and thus ensuring an order and procedural fairness on negotiations about shared waters (Ray 2014). One Bangladeshi scholar (Anonymous 3, 2014) admitted that it is only through the presence and interest of international actors that Bangladesh has been able to negotiate at all; third parties help keep India accountable.

Others in the region see international interventions in the Ganges–Brahmaputra problemshed as necessary and beneficial, albeit flawed. One prominent Indian analyst (Anonymous 6, 2014) has called SAWI the region's 'biggest talk shop' but has also admitted that it has been successful at instigating a much-needed conversation about transboundary water governance. Dialogue is necessary, but only the first step and it is difficult to quantify its effect on overall increased levels of cooperation throughout the problemshed. This, in turn, limits the effectiveness of international actors (Prasai 2014). Likewise, the fact that SAWI is creating a multilateral forum for the discussion of shared water issues is welcome, but undermined by India's lack of interest in multilateralism (Thakkar 2014). Thakkar (2014) argues that although it is good that SAWI is creating a multilateral forum (especially one that includes China) for the discussion of transboundary water issues, it is not able to get at the heart of the vested interests of the various stakeholders. In short, SAWI and the other international interventions are seen by many in the Ganges–Brahmaputra problemshed not as fundamentally flawed or a waste of time, but rather as having only a limited utility or potential for success because of the very nature of transboundary water cooperation (Anonymous 5, 2014).

In summary, international actors should not expect radical improvement in transboundary water interaction to happen rapidly (Anonymous 4, 2014), but it is part of the slow, steady and largely uncoordinated work that constitutes what one analyst calls 'chipping away' at the problem (Prasai 2014). As such, the international interventions in the management of the rivers of the Ganges–Brahmaputra problemshed, irrespective of how they are perceived and received, are part of the nudge governance of transboundary waters.

India has not been an enthusiastic participant in the Abu Dhabi Dialogue

International organisations and foreign aid donors may be making an overall positive contribution to transboundary water governance in the Ganges–Brahmaputra problemshed, but this contribution is weakened by the lukewarm support of regional powerbrokers. SAWI, for example, has had middling support from the South Asian states, and lacklustre support at best from the hydrohegemon, India. While SAWI has to a certain extent forged good relationships within the Indian central government, it has not been able to develop close relations with other important stakeholders in the system, such as state-level politicians and others who may have the influence to spearhead or stymie water governance projects (Anonymous 1, 2014).

One factor in the low buy-in of stakeholders is that the riparian states do not contribute funds to SAWI; only Australia, the United Kingdom and Norway do (World Bank 2014b, pp. 30–34). Host countries are often asked to contribute financially to international development projects to ensure their commitment. This has not been the case here. Asthana and Shukla (2014, p. 344) argue that this is a lost opportunity, as the collective investment in a regional water governance organisation could create a neutral space and unbiased environment away from the historical and regional biases and thus result in better regional strategies and transboundary water cooperation.

Part of the inability to engage with the most powerful actors in the system is a consequence of the poor reputation that the World Bank has within India (Anonymous 5, 2014). This distrust also encompasses SAWI, thus rendering the initiative largely ineffective in India (Anonymous 5, 2014). India has also been distrustful of the Abu Dhabi Dialogue, out of which SAWI was created. As discussed above, the ADD was started in 2006 and brings together senior members of government, academia and civil society from Afghanistan, Bangladesh, Bhutan, China, India, Nepal and Pakistan in a non-formal, consultative process intended to foster regional cooperation on various challenges relating to the Himalayan rivers (SAWI 2015c).

Because regional water management is a sensitive topic among member countries, the Abu Dhabi Dialogue has attempted to provide a neutral forum for open conversations among participants, and statements are non-attributable and there is no requirement for a consensus outcome (SAWI 2015c). Nonetheless, India did not want to participate in the Abu Dhabi Dialogue, but without the hydrohegemon, the process would have been greatly weakened (Prasai & Surie 2014). India did, in the end, participate in all six meetings of the Dialogue, but manifested its lack of commitment by sending a different, low-level representative each time, thus degrading the high-level intent of the process and undermining the continuity required for the success of these sensitive meetings (Anonymous 7, 2014).

India is not strongly committed to multilateralism in water governance issues in the Ganges–Brahmaputra problemshed – despite Prime Minister Modi's promising rhetoric in this regard – and this undermines SAWI's primary function of establishing a multilateral forum for transboundary water cooperation (Thakkar 2014). Thakkar (2014) argues that SAWI will remain ineffective until India truly commits to multilateralism. Earle et al. (2015, pp. 41–42) agree: 'India's refusal to take up any river management issue in a multilateral forum reduces the scope for possibility of any basin-based organisational framework to emerge.'

But India's middling commitment to multilateral processes may also be undermining its hegemony in South Asia. As Jones (2008, pp. 183–193) argues, India has global ambitions, but achieving them requires India to take a lead role in stabilising its own region before it will be accepted by other great powers as a member of the club. In light of Jones' analysis, it can be argued that India's reluctance to embrace the Abu Dhabi Dialogue and SAWI, and to forge a leadership role for itself within these regional water governance processes, not only weakens foreign-led interventions, but may also be a missed opportunity for India to strengthen its leadership position and profile internationally.

The hydro-hegemon is also not strongly committed to the idea of transboundary water cooperation. India's National Water Policy illustrates that the values espoused in the Abu Dhabi Dialogue process have not been embraced by India's hydrocrats and ruling elite. The National Water Policy was adopted in 2012, the same year that SAWI Phase II commenced, and the year of the sixth and final Abu Dhabi Dialogue. Yet, the Indian National Water Policy does not reflect any of the language prevalent in the discourses around water governance in South Asia at that time, that is, the language of deliberative governance, multilateral cooperation, regional stability, sustainable development and so on. Of the policy's total of 16 articles, only one is devoted to transboundary waters. It consists of two short clauses and stipulates a strictly bilateral approach to transboundary water governance rather than a multilateral one. The policy is clear: 'keeping paramount the national interest' is key in any negotiation over transboundary waters (Government of India 2012, Article 13). This is not the rhetoric of benefit sharing and positive-sum outcomes. And although the policy stipulates that India's water resources are to be managed as a common pool resource, there is no mention of deliberative governance, or other collaborative approaches to governance. Climate change is mentioned in the National Water Policy, but not climate change resilience – another focus area for foreign interventions in the Ganges–Brahmaputra problemshed.

In summary, India is neither enthusiastic about the idea of transboundary water cooperation through deliberative or multilateral approaches, nor is it an active supporter of international actors that pursue this agenda in the Ganges–Brahmaputra problemshed. This curtails the ability of third parties to facilitate transboundary water cooperation as the hydro-hegemon's active participation in or, at the very least, tacit support for, such initiatives confers legitimacy on them. As Asthana and Shukla (2014, p. 344) suggest, India's hydro-hegemony makes 'multilateralism or third-party interventions in a regional waterscape … not only difficult but also impossible'.

Rejection of the Ganges Strategic Basin Assessment undermines international efforts

A further illustration of the weak support of riparian states for international organisations and aid donors engaging in the transboundary water governance of the Ganges–Brahmaputra problemshed lies in the reaction to the Ganges Strategic Basin Assessment. India, Nepal and Bangladesh initially rejected this World Bank-led study that tried to collaboratively establish the scientific basis for possible benefit-sharing arrangements within the Ganges basin.

In 2009, an international, inter-disciplinary team led by World Bank economists commenced a multi-sectoral study of the Ganges basin (Sadoff 2009). Up until then, the World Bank points out, there had been no such analytical framework that could be used by riparian states for cooperative planning in the Ganges. Information and data are surprisingly scarce and difficult to obtain in South Asia (World Bank 2014a, p. xiii).

The conceptual underpinnings of the Ganges Strategic Basin Assessment are rooted in the idea that transboundary water cooperation can be improved through benefit sharing. However, benefit sharing is a negotiation tactic (one in which parties to the negotiation come to see the contested issue as an opportunity for creating positive-sum outcomes – arrangements with many winners – rather than zero-sum outcomes – a winner-takes-all scenario). Benefit sharing is not a method of conflict resolution, nor a means of creating positive transboundary water interaction. It may, however, come about that within positive transboundary water interaction, benefit sharing negotiations are pursued because the parties are already willing to cooperate and trust each other. In other words, benefit sharing may be a consequence of cooperation rather than its cause. But this nuanced understanding of transboundary water cooperation is not explicitly present in the Ganges Strategic Basin Assessment.

The findings of the Ganges Strategic Basin Assessment answer ten fundamental questions relating to some commonly held assumptions about the transboundary water governance of the Ganges basin (World Bank 2014a, p. xiv). Significantly, the report dispels several myths about the hydrology of the basin and challenges the perceptions of how to manage shared water problems, especially flooding, water storage and hydropower development (Sadoff et al. 2011). Specifically, '[h]ydropower development and trade are confirmed to hold real promise' at the same time as 'the findings of this study refute the broadly held view that upstream water storage (i.e., reservoirs) in Nepal can control basin-wide flooding' (World Bank 2014a, p. xiv). The findings of the Ganges Strategic Basin Assessment have, among other outcomes, influenced the design of World Bank-supported investments in flood management and agricultural water management in northern India as well as climate resilience programmes in Nepal (Rollason 2012).

Why, then, was the report widely rejected by stakeholders? The intent of the Ganges Strategic Basin Assessment was, after all to instigate collaborative, evidence-based, problem solving. The World Bank team began with the conventional wisdom, which is that large dams are needed in Nepal to stop flooding and provide water for irrigation throughout the basin. The assumption that the World Bank wanted to challenge was that the entire system, all the way into Bangladesh, could be regulated with upstream infrastructure. The Ganges Strategic Basin Assessment was to be an antidote to 'so much contentious history over treaties' (Colopy 2012, p. 320).

Yet, when a draft of the report was released in mid-2011, instead of instigating discussions between riparians about benefit sharing options, it was considered controversial and was widely rejected (Siddiqui 2014). India rejected the Ganges Strategic Basin Assessment on the grounds that the report did not use any official Indian data – a disingenuous argument considering that India's government did not share any data with the World Bank in the first place (Siddiqui 2014). Maintaining secrecy around hydrological data offers India the option of rejecting or otherwise undermining the validity of social or environmental impact assessments that do not use the 'correct' data that the authors of such assessments have been denied access to (Anonymous 9, 2014).

The fact of the Ganges Strategic Basin Assessment being weak on Indian data was also of concern to Nepal. Among other objections, Nepal noted that 'data from Indian side has not been used to that extent as it should be' and questioned 'the reliability of the model as the model considers the data of Nepal and Bangladesh only' (Jalsrot Vikas Sanstha 2012, pp. 1–2). Many other concerns were raised in an official rejection of the Ganges Strategic Basin Assessment, which provided comprehensive point-by-point counterarguments to the report (Siddiqui 2014). Furthermore, a consultative workshop was organised in Kathmandu in mid-2012 by Jalsrot Vikas Sanstha (a Nepalese NGO that promotes IWRM throughout the region) and the Global Water Partnership, Nepal. The consultative workshop aired concerns about 'such type of misleading report' and raised objections to 'the attempt of World Bank to undermine Nepal's contribution in the Ganges basin' (Jalsrot Vikas Sanstha 2012, pp. 1–2). The workshop's findings were submitted to the World Bank, which in turn attempted to turn the critique of the Ganges Strategic Basin Assessment into an opportunity for further dialogue (Sayed 2012). Resentment, however, has lingered over the 'Indo-centric' report, and the sentiment remains that 'Nepal … has every right to say NO to large projects that submerge for perpetuity vast tracts of fertile lands and villages in Nepal' and that 'Nepal, in the meantime, should implement projects that will cater to her own domestic water and energy needs' (Pun 2013, p. 12).

As for the third Ganges basin riparian, Bangladesh, it is of little concern to the Ganges Strategic Basin Assessment (which, according to some Bangladeshi analysts (Siddiqui 2014), may be symptomatic of a broader trend of foreign aid donors and international organisations paying less attention to Bangladesh than to India or Nepal). Bangladesh rejected the Ganges Strategic Basin Assessment unofficially, and at least one Bangladeshi contributor to the report (Anonymous

9, 2014) is now embarrassed to be associated with a report that has apparently misrepresented their work.

The rejection of the Ganges Strategic Basin Assessment was a problem for the World Bank because the report was to provide the justification for future infrastructure investment in the region (Siddiqui 2014). This, too, was a concern for India, Nepal and Bangladesh: the Ganges Strategic Basin Assessment puts forth solutions to transboundary water governance problems that are primarily about engineering, infrastructure or technology, and this is another major reason why the report was rejected (Anonymous 7, 2014).

Considering that the economic paradigm (Allan 2006, pp. 46–49) of river management is one that Indian, Nepalese and Bangladeshi hydrocrats largely favour (as discussed in Chapter 2), perhaps the underlying reason for the unanimous rejection of the Ganges Strategic Basin Assessment is not that it advocated predominantly scientific solutions to wicked problems of policy. Rather, the report was rejected because it implied those solutions should be, or could only be, implemented through external actors such as the World Bank, thus diminishing the agency of the riparian stakeholders themselves. The most effective solutions to international development problems tend to be the locally devised ones; no matter how sound an external solution may be, if it is driven by third parties it is likely to be resented or rejected, and thus rendered ineffective.

A further issue with the Ganges Strategic Basin Assessment is the manner in which it was conducted. The World Bank did not lead the study according to the principles of deliberative governance that the World Bank is advocating (through SAWI) for the management of the transboundary rivers in the Ganges–Brahmaputra problemshed. That is to say, the assessment was not a consultative process, nor was it based on wide-ranging and open discussions even though 'to promote an open, evidence-based dialogue' (World Bank 2014a, p. xiii) was one of the stated objectives of the report. The contributors were not able to see what the other contributors were writing (Anonymous 9, 2014). In short, the Ganges Strategic Basin Assessment was a failure because the study – and, indeed, SAWI – are perceived as top-down initiatives (Anonymous 9, 2014). As discussed above, the attempt to impose bottom-up, deliberative processes by third parties working in a top-down way is problematic and resented within South Asia (Anonymous 11, 2014).

The paradox of the Ganges Strategic Basin Assessment is that it was not intended as a top-down initiative but rather as an inclusive one, in line with the World Bank's recent attempts at taking a 'lighter touch' within larger coalitions and becoming 'more agile and effective as a participant in broad coalitions' (Evans et al. 2015, pp. 190–191, 200). Perhaps the experience of the less-than-inclusive process of developing the Ganges Strategic Basin Assessment will provide some lessons for a similar state-of-the-basin study of the Brahmaputra basin (World Bank 2014b, p. 16).

The Ganges Strategic Basin Assessment was ultimately published by the World Bank in 2014. It is one of the most thorough and up to date reports on the hydrological, economic and social state of the Ganges basin. The reputational damage, however, has been done to SAWI and this will likely affect its future efficacy. The World Bank's position as the champion of deliberative governance in the transboundary water governance of the Ganges–Brahmaputra problemshed has been undermined by the less-than-deliberative development of Ganges Strategic Basement Assessment.

A role for international actors?

Are the normative aims of international organisations and aid donors likely to be achieved in the Ganges–Brahmaputra problemshed? Based on the insights and experiences of riparian stakeholders, international actors engaged in increasing transboundary water cooperation in the Ganges–Brahmaputra problemshed are unlikely to have a direct and significant effect in the short term on improving water interaction between India, Nepal, Bhutan and Bangladesh. They are but a drop in the ocean in the complex hydropolitics of the region – but that is not to say that they are unwanted, unnecessary or making matters worse. After all, every drop counts. It is just as difficult to attribute failure of governance as it is to attribute success (Pahl-Wostl 2015, p. 32).

Instead, the more pressing question is how all parties – those from within the problemshed, the elites of riparian states, stakeholders that are dispossessed, as well as interested third parties – can or should better address the complex environmental, security, and policy challenges of today and tomorrow. Swain (2011, p. 698) points out that the urgency 'to meet the new challenges that might arise from climate change' is increasing. In light of these complex new challenges, Verady et al. (2014, p. 28) argue that hydro-diplomacy will be increasingly needed to anticipate, mitigate and resolve conflicting claims on shared water resources in transboundary regions. International actors are likely to be called upon more frequently to facilitate dialogue between states trying to balance national interests with regional welfare and economic security. In finding the right balance, argue Verady et al. (2014, p. 28) 'lies the great challenge of hydro-diplomacy and the concomitant potential of hydro-solidarity'.

8 Obstacles to positive water interaction in the Ganges–Brahmaputra problemshed

Hydro-diplomacy is not widespread in the Ganges–Brahmaputra problemshed and there is not much hydro-solidarity. This makes the work of international actors in this region more challenging – and all the more important. There are several other factors that likewise make the task of increasing transboundary water cooperation more difficult and more pressing. Some of these are discussed in this chapter.

The first obstacle is the still nascent civil society. The second obstacle is the weakness of a regional identity within South Asia, which means there is no effective regional body for governing social and economic issues, such as the development of water resources. Furthermore, there is a lack of trust between riparians and a zero-sum attitude to water sharing prevails. This is compounded by the securitisation of water issues in the region. A further and related factor is the culture of secrecy that surrounds hydrological data. These five obstacles are discussed in turn.

Underdeveloped civil society

South Asia is one of the most poorly governed regions in the world, with inefficient and inequitable deployment of resources, crippling debt burden, social divisions along ethnic and sectarian lines, as well as rampant corruption and vulnerability of civil society organisations (Baqai 2011, p. 27). As Baqai (2011, p. 31) notes:

> [t]he South Asian state is both strong and weak. On one hand, a high degree of 'state-ness' is witnessed, while on the other hand, the breakdown of state authority, as a result of social disorder makes political upheaval the norm.

In addition to being 'soft states' (that is, ones unable to make citizens obey laws and regulations, or impose discipline over them), the states of South Asia are still defining themselves as nations. South Asia is in the process of nation-building, and the region has been described as one of 'nations in hope' but not in being (Baqai 2011, pp. 21–22, 30). To further the process of nation-building, or to compensate for it not being far enough advanced, there is a dominant nationalist discourse in the region that accords precedence to nations and nationalism over the region (Behera 2002, p. 210). Prasai and Surie (2013, p. 27) see this nationalist discourse as affecting transboundary water too. One of the characteristics that they identify as defining water governance in South Asia is that it is nationalistic in terms of discourse.

With nationalistic discourses dominating, political leadership in South Asian countries has been unable, according to Behera (2002, p. 210), to imagine and evolve a mind-set that could be truly characterised as 'South Asian'. The task of creating a regional mind, then, has fallen to civil society. Unfettered by the nation-state ideology, the players in the civil society are better placed to conceive, shape and nurture the idea of 'South Asia' (Behera 2002, p. 210).

Civil society – meaning the network of political, economic and social institutions, and interest groups such as political parties, religious and cultural associations and so on – is critical to governance in general and can further democracy as well as 'safeguard individual rights from any kind of authoritarianism' (Baqai 2011, p. 24). In South Asia, the rapid growth of NGOs has opened up new communication channels at the grassroots level. These support a broad spectrum of issues ranging from gender issues and human-rights violations to ecology. They play an active role in civic mobilisation and public advocacy and in organising activities across national boundaries in their areas (Behera 2002, p. 212). Civil society has also been key in establishing the beginnings of the idea of 'South Asia' in the absence of a regionalist outlook in the polity. Unfettered by the nation-state ideology, the players in the civil society are, according to Behera (2002, p. 210), better placed to conceive, shape and nurture the idea of 'South Asia'. At the same time, in South Asia there is mutual hostility between civil society and the state (Baqai 2011, p. 25).

Civil society has not been particularly active in addressing environmental concerns. Environmental concerns are widespread in South Asia, according to Iyer (2007, p. 195), but they are not equally strong or influential in all countries. There are some powerful environmental movements, but they are not very effective and cannot be said to be mainstream. Historically, environmental movements in South Asia have been linked to anti-dam movements, but as Whitehead shows this has had mixed results in terms of bolstering domestic environmentalism. Transnational linkages in the anti-Narmada dams movement was, in some ways, a double-edged sword. It produced a 'middle ground' of international environmentalism that gained strength through its international constituency. But, as Whitehead (2008, p. 297) argues, it also suppressed emerging local currents, contradictions and dynamics, particularly in its use and representation of indigenous peoples.

The role of civil society has also been limited in transboundary water governance in South Asia. As Hill (2013, pp. 255–256) shows, NGOs in the region tend to be preoccupied with donor-related developmentalism, rather than advocacy. And they are constrained in the extent to which they can express dissent. Part of the challenge in realising the potential of these networks, writes Hill, is the politically fragmented nature of the subcontinent. It is very difficult for pan-South Asian networks to regularly meet and forthrightly take up issues associated with transboundary water resources (Hill 2013, pp. 255–256).

Despite these challenges, there are alliances beginning to form between civil society groups across the region. Examples include the South Asian Solidarity on Rivers and Peoples (SARP), Narmada Bachao Andolan (NBA), and the South Asian Network for Dams, Rivers and Peoples (SANDRP), while Imagine New South Asia (INSA) is an example of a new breed of regional think tanks that look at, among other things, transboundary water issues. Hill (2013, pp. 255–256) argues that 'these are fairly nascent and, as yet, their influence on broader policy debates is hard to discern'. Chokkakula (2012, p. 45) suggests that one effect of '[t]he multi-actor driven multi-scalar everyday politics in the transboundary spaces of interstate water disputes' is that they 'de-centre' the state – they diffuse the drivers of action away from state-actors. Nonetheless, as discussed below, the water governance in South Asia remains on whole state-centric and inter-governmental.

There is also a growing network of national and international universities, think tanks, NGOs, public commentators and so on working on and influencing the resolutions of transboundary water issues. The salient point here is that there are many different actors involved in transboundary water governance in the Ganges–Brahmaputra problemshed, and not all are state actors. They collaborate across borders, build trust and so on and it is part of hydro-diplomacy to consider how these informal dynamics can be harnessed in formal and informal settings (Keskinen et al. 2014, p. 38).

Nonetheless, in South Asia the state-centric narrative still dominates hydropolitics – and the development of hydropower in particular – despite this emerging network of non-state actors in water governance. Hill shows that there have been some hopeful developments in terms of the potential of civil society and people's movements. As mentioned above, this is especially so in India because of its extensive and long-running networks contesting the terms under which hydropower development takes place. However, most states, including India, continue to be highly centralised in terms of their water bureaucracies and it is difficult for these groups to have a significant effect on water governance. In the smaller nations, Hill (2015, p. 741) points out that NGOs are predominately focused upon service delivery and have limited resources or capacity to devote to advocacy or political mobilisation.

Advocacy groups in particular, argues Hill (2015, p. 741), are constrained in how they can express dissent. These difficulties faced by civil society groups are amplified when it comes to transboundary networks in South Asia, and so despite several promising attempts, there are few genuinely regional initiatives, or even those that successfully operate across international boundaries (Hill 2015, p. 741). That, in turn, limits good water governance in the problemshed.

Weak regionalism, strong state-centrism

Partly due to the weakness of civil society and non-state groups, water governance in the Ganges–Brahmaputra problemshed continues to be seen – both by riparian stakeholders and by international actors – primarily as an inter-governmental undertaking. That is incompatible with what Iyer (2010, p. 76) describes as 'a consultative, interactive approach to conflict-resolution' that also involves those who have a stake in the decision, namely the farmers, industrial establishments, municipalities and people in general. Everyone, in other words, is a water-user. Iyer's ideal is far from the reality of water governance in the region. Prasai and Surie (2013, p. 27) argue that water governance policies in South Asia are characterised by three features. They are: 1) statist in terms of decision-making; 2) nationalistic in terms of discourse; and 3) reductionist in terms of scope.

The characterisation of water governance in South Asia as being statist and nationalistic stems, in part, from a weak regional identity in South Asia. Regionalism in a political, economic and national security sense may be a long way off yet. But it is considered desirable by international organisations and foreign aid donors. One normative claim of international actors engaging in transboundary water governance in the Ganges–Brahmaputra problemshed is greater regionalism. Why is regionalism desirable for transboundary water interactions? As Hill (2015, p. 732) explains, regional integration is assumed to lower the cost of trade and increase market access. This is purported to be particularly beneficial for economically marginalised people living in remote and landlocked areas. In addition, efficient and sustainable technologies will replace and augment previous ways of doing things. These flows of expertise, investment and technology, argues Hill (2015, p. 732), will all have the benefit of reducing poverty and the likelihood of conflict.

Dash (2008, p. 110) sees four major challenges to the development of true regionalism: India's hegemonic status and Pakistan's relentless challenging of that position; the existence of weak ruling coalitions; ethnic crises; the nuclear issue. Regionalism in the truest sense is unlikely any time soon in South Asia, but that should not be a surprise or a worry; Jain (2010, p. 18) notes that it took decades for a free-trade and a common market regime to be established in Europe. Furthermore, as Jones (2008, p. 183) points out, every region in the world is unique, and purist notions of what constitutes a regional community are unlikely to be relevant to South Asia. Indeed, while South Asia is poorly integrated, it is also, in Kugelman's (2012, pp. 178–179) words, 'linked together by water co-dependencies' and this paradox is at the heart of transboundary water arrangements in the region.

Despite the high level of interdependence over water issues, the Ganges remains one of the few large international basins in the world without a permanent institutional mechanism that involves multiple riparians (World Bank 2014a, p. 32). The same is true for the Brahmaputra. Despite years of discussions and negotiations, there are no basin-wide mechanisms for communication and cooperation. Instead, bilateral arrangements (such as the

Indo-Bangladesh Joint Rivers Commission, or the Indo-Nepal Joint Committee on Water Resources) continue to be the preferred mode of cooperation within the Ganges–Brahmaputra problemshed (World Bank 2014a, p. 27). India's role in establishing and maintaining strictly bilateral arrangements is examined in Chapter 9.

There is one major regional organisation, the South Asian Association for Regional Cooperation (SAARC) but, as Kugleman (2012, pp. 178–179) argues, it 'is not nearly as dynamic as regional grouping like the European Union or the Association of Southeast Asian Nations (ASEAN)'. The formation of the SAARC in 1985 was preceded by eight years of negotiations, but since its inception it has experienced only limited growth in terms of institutional development and programme implementation (Dash 2008, p. 3). Moreover, most SAARC activities:

> are confined to 'soft' areas of cooperation and to the holding of seminars, workshops, and short training programs. These activities may be useful, but they do not address priority core areas of trade and security and therefore lack visibility and regional focus.
>
> (Dash 2008, p. 96)

Transboundary water issues are not high on the SAARC agenda. As Hill (2008, p. 73) points out, 'water-sharing negotiations are conspicuously absent from discussions at SAARC, the regional body that would otherwise have seemed an obvious forum for such multilateral coordination'. Yet, SAARC was not originally mandated to address regional water issues, but in the context of global action on climate change it has begun to consider water-related problems. It has also created regional institutions for disaster management and meteorological research such as the SAARC Disaster Management Centre, and the SAARC Meteorological Centre, which promotes collective research in meteorology and weather forecasting in the region (World Bank 2014a, p. 27). There are also numerous multilateral dialogues as part of the SAARC process. Yet, according to Behera (2002, p. 219), there is an assumption that the governmental channels of the SAARC process have often failed to deliver and that it needs 'outside' support in terms of ideas, processes and mechanisms from non-official players.

Other reasons for the SAARC's middling effectiveness have been identified by Dash (2008, p. 120) as being 'India's hegemony, the lack of trust among South Asian states, and Pakistan's continuing challenge to India's domination in South Asia'. It is perhaps not surprising that other South Asian countries are apprehensive about what they perceive to be India's hegemony (Bhasin 2008, pp. 1–2). As Asthana and Shukla (2014, p. 242) point out, '[t]he relations between small and weak riparian states and their powerful neighbour [India] continue to be affected by suspicion, mistrust, envy, and domestic political concerns driven by the electoral politics of the respective states'. This general atmosphere of mistrust in the region is in addition to the apprehension about India's hegemony (Singh 2008, pp. 10, 35). This too, Dash (2008, pp. 172–174) notes, contributes to the very weak regional solidarity and identity with 'South Asia' in the region.

Weak regionalism is a factor in the low levels of political will for collective action on shared water issues. Despite numerous items on the transboundary water governance agenda, there is little political will among the riparians of the Ganges–Brahmaputra problemshed to address these. This limits the potential for increased transboundary water cooperation. As Earle et al. (2015, pp. 41–42) show, 'there is certainly a lack of political will in the basin, particularly in India, to provide the space for multilateral river institutions to effectively emerge'. Multilateral approaches are limited by the state-centrism and nationalistic discourses that Prasai and Surie (2013, p. 27) argue characterise water governance policies in South Asia.

One of the results of the weak solidarity and regional identity in South Asia is that the region lacks a security community. A security community means that a group of states has:

> achieved such a level of cooperation, or even integration that they simply do not consider fighting each other as a realistic possibility to resolve disputes, and stop preparing to do so (although such states may well continue to prepare to fight others).
>
> (Jones 2008, pp. 184–185)

Jones (2008, pp. 183–193) argues that this model of a security community is not currently applicable to South Asia, and is unlikely to develop in the foreseeable future. At the same time, he points out, some positive dynamics are occurring, especially in regard to India adopting a 'strategic altruism' (that is, considering the concerns of neighbours in its own security actions, especially for the purpose of fostering closer relationships) (Jones 2008, pp. 183–193).

Zero-sum attitudes to water sharing

At the root of the weak regional solidarity in South Asia, and resentment of India's hydro-hegemony, lies the lack of trust between states in the region. As Uprety (2014) points out, collective multilateral attention to the problems of transboundary waters in the Ganges–Brahmaputra problemshed has been rare because of the prevailing mutual distrust between riparians. This distrust means that 'the strategic approach of most of the countries is merely to theoretically engage in water-related initiatives, but practically advance only those serving their own specific interests' (Uprety 2014).

A further consequence of distrust between riparians is that a zero-sum attitude to water sharing prevails. A zero-sum view of transboundary water resources presumes that the water can only be used once, and only by one party. 'Not surprisingly,' argue Islam and Susskind (2013, pp. 134–135), 'it yields outcomes in which one side "wins" and others "lose"'. In contrast, a positive-sum view enables win–win arrangements, but it is predicated on a shared interest between riparian states that requires cooperation (Julien 2012, p. 51). Positive-sum approaches to water sharing are generally about optimising benefits for all riparians rather than dividing water (Qaddumi 2008, p. 1).

A zero-sum attitude to water resource dominates in Himalayan Asia, claim Wirsing et al. (2013, p. 13), that in turn is fuelling already intense geopolitical rivalry among riparian states. Chellaney (2013, p. 2) too sees a zero-sum mentality towards water as fuelling the rivalry and water nationalism in the region. That mentality is not unique to the transboundary scale; it is also apparent on the domestic scale. In India, for example, inter-state water disputes are subject to zero-sum thinking. As Paranjpye (2010, p. 124) argues, 'the lack of integrative thinking among all parties concerned' explains in part the failure of arbitration of these domestic water disputes.

The zero-sum, 'winner takes all' approach to water management allows for the securitisation of transboundary waters by establishing an adversarial dynamic between riparian states. Asthana and Shukla (2014, pp. 275–277) note that the securitisation of water means '[w]ater is treated as a geostrategic military tool instead of being elevated as a global common and a common resource recognised as vital for life'. The most prominent illustration of this is China, which is popularly perceived within South Asia as using its rivers as a 'weapon' and thus, according to Hangzo (2012, p. 2), treating them 'as a means of asserting control over its lower riparian neighbours'. This is a particularly powerful narrative in light of China's dam-building plans for the Great Bend of the Brahmaputra River, which alarm Wirsing et al. (2013, pp. 102–103) and Chellaney (2011). Water, along with other non-traditional security threats, is, according to Asthana and Shukla (2014, pp. 275–277), securitised across borders because of a bitter past, mistrust and hatred that dominates the socio-political structure in these countries.

At the same time, the prevalence of non-traditional security threats also presents an opportunity for increased regionalism and cooperation across different levels. As Hill (2015, p. 740) argues, conceptualising non-traditional security in a way that incorporates a nexus of food, water and energy necessitates the regional integration of upstream and downstream communities in the Himalayas in ways that achieve integration across many ecosystems. But the increasing emphasis on addressing non-traditional security threats, especially through the traditional security apparatus, can also lead to the securitisation of water.

Securitisation of transboundary waters

The securitisation of water fosters conflict, and thus undermines transboundary water cooperation. Countries, argues Chellaney, cooperate over transboundary water resources only insofar as it serves their national interest. But when a nation pushes a sense of grievance against a riparian neighbour, or when the transboundary water competition over economic development becomes fierce, then the potential for conflict grows (Chellaney 2013, p. 55). Prasai and Surie (2015, p. 31) understand that water is an emotive subject. Excessive nationalism undermines reasonable compromise on shared water issues the world over, not just in South Asia.

Understanding the riparians' vested interests is, according to Mirumachi (2013, p. 9), important in understanding when and why a state may securitise water issues. Furthermore, Mirumachi and Allan (2007, p. 4) show that conflict intensity over transboundary waters increases as the perception of the issues by the state changes. In other words, as issues become perceived as more of threat to the state, they are prioritised in the national agenda. Thus, they receive more attention and attract allocations of various state resources. They are politicised.

Non-politicised issues are not ones for state action, and not included in public debate. Politicised issues are managed within the standard political system, and are part of the public policy debate. An issue is securitised when it requires (or is perceived to require) emergency actions beyond the state's standard political procedures (Emmers 2007, p. 111). Significantly, a securitised issue can be intensified, claims Williams (2003, p. 516), 'to the point where it is presented and accepted as an "existential threat"'. That existential or extraordinary threat may, according to Warner and Meissner (2008, p. 258), legitimise extraordinary measures, thus elevating a non-security issue to a security issue through a 'speech act'. Speech acts can also lead to the opposite process, that of de-securitisation, which moves issues off the 'security' agenda and back into the realm of public political discourse and 'normal' political dispute and accommodation (Williams 2003, p. 523). One example of this is Apartheid in South Africa moving from a security issue to a generally political one (Emmers 2007, p. 111).

What, then, does the securitisation and de-securitisation of transboundary waters look like? Except in a realist understanding of international relations (Julien 2012, p. 59) water is not often the mainstay of a state's security. Yet, transboundary rivers do involve security and risk concerns for both the environment and humans, and can be the focal point of a broader conflict (Warner & Meissner 2008, p. 254). Crucial to an understanding of the securitisation of water is, according to Mirumachi (2013, p. 9), an understanding of the hydropolitical context because it will help explain the vested interests that exist between various actors.

Mirumachi (2013, p. 1) shows that securitisation theory examines not only the physical threats to and from the environment, but also the threats based on imagined and discursive constructions. Environmental security thus asks 'who securitises what and how' through an examination of narratives and discourses. Warner and Meissner (2008, p. 258) add that because all is fair in political struggle, it is difficult to tell whether securitisation is a genuine cry of despair or a cunning strategy. There is, in other words, no objective measure of water security.

Water security is, according to Julien (2012, p. 59), socially constructed: 'To politicise, to securitise or to violise an issue is always a collective choice, conscious or not, rather than a necessity.' Securitisation results from a successful manoeuvre by socially influential actors to convince the relevant audiences that something represents an existential threat to some cherished value. Extraordinary measures are then justified to react to this threat.

Julien's point about the influence of actors is particularly salient for the securitisation of transboundary water issues in South Asia. It is the hegemon's prerogative to securitise an issue and '[n]on-hegemons have much less ability to de-politicise a matter that a hegemon considers an issue, and environmental agendas set by the powerful inevitably lead to non-hegemons securitising the resource' (Zeitoun et al. 2013, p. 2). This is the case in the Ganges–Brahmaputra problemshed.

Transboundary water security is an issue on the political agenda of India, Nepal, Bhutan and Bangladesh. In particular, the energy and resource security aspects of water have, according to Wirsing et al. (2013, p. 10), risen to a position of great importance on Asian strategic agendas and are fuelling geopolitical rivalries. At the same time, water itself is an emotive subject, and Prasai and Surie (2015, p. 31) show that '[d]rumming up of nationalistic fervour on water is politically expedient, particularly when securitising water becomes a strategic intent and governments in South Asia (as elsewhere) routinely indulge in that enterprise'.

One example of the securitisation of transboundary waters in the Ganges–Brahmaputra problemshed is that of interaction between India and Nepal over the Tanakpur agreement. In 1983, India began construction on a barrage close to the border with western Nepal. The initial construction was wholly on Indian territory, but India failed to consult Nepal (Subedi 2005, p. 128). By the 1990s it become clear that to maximise the objectives of the barrage – regulated flows to irrigate 1.61 million hectares of land downstream, as well as run-of-the-river hydropower production, both of which would benefit India only – some construction in Nepali territory was needed (Mirumachi 2013, pp. 4–7).

To create the left afflux bund (or the eastern wall to hold back water), India request 577 metres of Nepali land, which in effect would also submerge 2.9 hectares of land in Nepal (Singh 2008, p. 24). This request was, according to Mirumachi (2013, p. 7), essentially a speech act that securitised the shared waters. The letter of May 1991 from the Indian Prime Minister to the Nepali head of state took a strong position on the need for hydraulic infrastructure. Mirumachi (2013, p. 7) argues that the letter clearly emphasises that India and Nepal are both at the mercy of nature, with monsoon rains bringing about potential disaster. The 'permanent solution' to this threat is presented as human intervention in the form of engineering.

Thus India constructed an existential threat, then presented a seemingly obvious and unalterable solution to it. Mirumachi (2013, pp. 9–10) continues that India seized the opportunity to unilaterally construct the legal status of the Tanakpur agreement even while it was still being debated in Nepal. She argues securitisation was possible because of the power asymmetry between the two countries. India's securitisation of the issue produced a situation where the development of the barrage could not be reversed, as it was now officially recognised in the legally binding Mahakali Treaty. Moreover, this basin-wide agreement was limited in promoting equitable water allocation despite clarifying many governance aspects of the river, including the Tanakpur Barrage (Mirumachi 2013, pp. 9–10).

There are other examples of how transboundary waters are securitised in the Ganges–Brahmaputra problemshed. As discussed below, the secrecy around water related knowledge securitises water because it elevates it into the realm of national security even though, as Prasai and Surie (2014) have said, hydrological data is of no use to military strategists. Nonetheless, influential commentators such as Brahma Chellaney (2011) continue to fear 'water wars' in the region. Baqai (2011, p. 88) shows that the Indian media in particular favours war-terminology, especially in coverage of water issues with Pakistan, in its frequent use of phrases such as 'battle for peace', 'arch enemies', 'ancient feud' and so on. Local media too tends to present transboundary water issues in an alarmist way, often blaming upstream or downstream communities for water related disasters such as flooding (Singh 2008, p. 26).

Once water is securitised, it can have detrimental effects on the production of knowledge around transboundary water issues. For example, individual researchers, or those not affiliated with an Indian government institution, are often restricted in where they can travel to within India for their research, especially in the north-eastern part of the country – the mountainous region with significant hydropower potential and on-going insurgent activity (Mitra 2010, p. 19). Some scholars have even been strongly advised against doing research on transboundary waters by India's security agencies. There is also scepticism and suspicion from the Arunachal Pradesh government towards research, NGOs and others wanting to work on or research water-related issues because it fears any negative outcome may hurt the state's economic interests (Bhonsale 2014). Thus the securitisation of water is not only a rhetorical or abstract phenomenon. It has real and profound effects on water interaction.

Water gets securitised in many ways and by many actors. As Hill (2015, p. 733) shows, 'social construction of this threat emanates from many sources, including security intellectuals, religious extremists, circulation savvy media houses and popular representations'. In relation to the Hindu Kush Himalayan region especially, he writes that because water is securitised, accelerating the construction of a hydropower dam is portrayed as something akin to a geo-strategic imperative. This has the effect that some commentators demand the silencing of dissent and the scrapping of due process around the social, environmental and economic impacts of planned projects within their borders (Hill 2015, p. 733). The consequence of this, argues Hill (2015, p. 729), is that 'the ongoing securitisation of water constrains the terms of debate under the guise of a unified national interest and enables large-scale dams to be constructed without due process'. The remedy he calls for is the de-securitisation of transboundary waters in the region (Hill 2015, p. 729), a point elaborated on in Chapter 12.

Secrecy around water knowledge

The securitisation of transboundary waters is also made possible because of the limited availability of accurate hydrological data in the region and a culture of secrecy that surrounds the knowledge that is available. A culture of secrecy and suspicion prevails across all governments in South Asia, not just the Indian one, and stifles any inclination to declassify or share data (Behera 2002, p. 227). This is a big problem for transboundary water governance. As Vaidya and Sharma (2014, p. 3) argue, '[k]nowledge gaps with respect to present and future hydrology pose a real constraint to water resource management' in the Ganges–Brahmaputra problemshed. There are a number of challenges that arise out of these knowledge gaps.

The first challenge is the culture of secrecy that not only hinders cooperation and collaboration, but could also create political tensions and lead to conflict. Chellaney (2013, p. 56), for example, sees the denial of hydrological data in a critically important season as tantamount to political leverage against a downriver state. As one Bangladeshi analyst pointed out, India is able to reject environmental impact assessments conducted in Bangladesh as incorrect on the grounds that such assessments do not use accurate upstream data – data that India refuses to share (Anonymous 9, 2014).

Why is the lack of water data a problem for transboundary interaction? As Prasai and Surie (2015, p. 31) show, in the absence of accurate, verifiable data that are shared across borders, riparian stakeholders (including the media, civil society organisations and communities) frequently end up using 'dubious information to accuse one another of diverting or exploiting transboundary water sources to the detriment of others'. For example, the Bangladeshi press has for a long time now accused India of diverting more than its fair share of water from the Ganges River through the Farakka Barrage. Bangladeshi political parties have expressed similar concerns over India's proposed diversion of Teesta River waters, and contentions have arisen frequently around implementation of the Kosi and Mahakali treaties in Nepali and Indian media (Prasai & Surie 2015, p. 31).

The limits on information about hydrology and irrigation withdrawals in India are largely a matter of policy, not availability. The Indian government has, for example, a sophisticated satellite information system for monitoring water resources (India-WRIS), but this database restricts public access to all hydrological data and information for India's northern transboundary rivers (that is the Indus, Ganges and the Brahmaputra) on the grounds that it is classified (Prasai & Surie 2015, p. 26). India's Central Water Commission also consistently rejects requests for information. Information such as sediment flow data or planning documents about barrages and other structures on transboundary rivers (the Kosi River, for example, a tributary of the Ganges) are classified in the interest of national security. Prasai and Surie (2015, p. 28) argue that this illustrates the highly securitised approach towards water and climate data in India. In comparison, in Bangladesh and Nepal general information pertaining to transboundary rivers is not considered classified or secret, but it is not collected, retained or disseminated in a consistent way.

India's policy of secrecy regarding transboundary hydrological data does not preclude other riparians from conducting their own research and hydrological monitoring – but they do not do so methodically (Ray 2014). That is a further example of the low political will in Nepal and Bangladesh for tackling transboundary water issues. India's neighbours could invest in their own hydrological studies, but generally do not do so. Instead, international actors have been taking the lead on knowledge production and dissemination in the problemshed. The International Centre for Integrated Mountain Development (ICIMOD) in particular has been conducting multilateral scientific research projects in the Hindu Kush Himalaya.

One feature of many transboundary hydropower projects in South Asia is that they were not known through government-to-government communication, but through public media; timely and adequate information is never easily or fully given, claims Singh (2008, p. 16). The lack of information sharing is not only a government-to-government phenomenon, but also a government-to-people one. As Behera (2002, p. 227) argues, in South Asia there is a distinct barrier between officials and the public; between those 'inside' the establishment, and those 'outside', that is, civil society. The two operate from fundamentally different information bases. The government information base remains too narrow and that of the non-government sources is wide but not well-informed. As a result, Behera (2002, p. 227) claims, there is considerable mutual suspicion rather than mutual interaction.

There have been recent attempts to break through this culture of secrecy. India's 2012 National Water Policy, for example, does promise the declassification of some hydrological data (Government of India 2012, paragraph 14.1). Yet, around the time that this policy was being drafted, India rejected the Ganges Strategic Basin Assessment in part on the grounds that the report did not, according to Siddiqui (2014), use any Indian government data – data that the Indian government was not able to share with the World Bank because it is classified. Indeed, the Ganges Strategic Basin Assessment notes that in particular, very little information is available on hydrology and irrigation withdrawals in India (World Bank 2014a, p. xiii).

The Indian Government's Hydro-Meteorological Data Dissemination Policy 2013 also outlines a procedure for public access to hydrological information on India's river systems, but disclosure or use of this information in its original format is strictly prohibited. Thus while formally there is a commitment to greater information disclosure, in practice there is weak compliance with the letter of law at a bureaucratic level (Prasai & Surie 2015, p. 26).

In addition to the prevailing culture of secrecy around hydrological data, there are several other obstacles to information sharing in the problemshed. In a recent Asia Foundation study, Prasai and Surie (2015, p. 29) identify these as: discrepancy between formal and informal disclosure of information; limited proactive disclosure; weak implementation of the Right to Information Acts; and poor data and records practices.

Prasai and Surie (2015, p. 28) have found that the government departments in Bangladesh and Nepal appeared more willing to provide water-related information 'informally' – on the phone or in an email – rather than through official Right to Information channels. They have also found that data can be sporadically accessed at the local level. Restrictions on data-sharing that exist at the national level, tend to occasionally fail at the local level, although information at the local level is limited and fragmented (Prasai & Surie 2015, p. 28).

The inconsistency with which information is disclosed – or not – is also highlighted by Molden (2014), who points out that difficulties may unexpectedly ease up over data sharing during crises or unforeseeable events such as large floods. At the same time, as Prasai and Surie (2015, p. 28) show, even in the aftermath of the 2008 floods in India's Bihar province and bordering Nepal, government agencies on either side of the India–Nepal border have made limited efforts to purposefully disclose relevant information to local communities, even on critical issues such as relief and rehabilitation efforts. This inhibits even politically neutral work, such as humanitarian and disaster relief efforts by non-government organisations (Surie 2014). Significantly, even when some information is available, there seems to be a gulf between what end-users need, and what they have access to (Prasai & Surie 2014).

The restrictions on access to information are, say Prasai and Surie (2015, pp. 28–29), easy to understand. The governments in India, Nepal and Bangladesh either lack the political will or have not developed the capability to administer managed or regulated access to hydrological data. The choice, then, is between open access and blanket restriction, and governments appear to find the latter approach easier to adopt. This preference, they argue, prevails over any internal and external clamouring for increased transparency (Prasai & Surie 2015, pp. 28–29).

The ease of imposing a blanket restriction on information sharing is facilitated by poor data and records management practices. Prasai and Surie (2015, p. 26) note that in many instances, while departments appeared willing to provide information, the information could either not be traced or was not always held by the department mandated to collect it. Where it was available, it was often found to be incomplete, difficult to interpret or existed in a format that was not user-friendly (Prasai & Surie 2015, p. 26).

The poor state of record-keeping, however, is not unique to water governance in the region (Prasai & Surie 2015, p. 26). But it is a particular problem for the water sector. The low quality or incomplete nature of information that is available impedes water management. Accurate and comprehensive data is a requirement of good water governance, especially planning and monitoring, and is thus, according to Molden (2014), a critical aspect of transboundary water cooperation; without thorough data exchange, there cannot be proper planning.

At the same time, information sharing is only one aspect of cooperation. Relationships, argues D'Souza (2014), are more important than data, no matter how comprehensive, and focusing excessively on the problems of sharing hydrological data between India, Nepal, Bhutan and Bangladesh is missing the point of what is the cause of discord between them over transboundary water issues. The problem of managing South Asia's river basin is not a question of inadequate river modelling data, it is a question of hydro-diplomacy (Anonymous 7, 2014).

Hydro-diplomacy, however, can be served through information and knowledge sharing. The act of exchanging data and bringing stakeholders together to negotiate what should be exchanged is in itself valuable insofar as it strengthens relationships and builds trust. Likewise, the river basin modelling projects that are being undertaken by international organisations such as the Australian Commonwealth Scientific and Industrial Research Organisation (CSIRO) have dual benefits. They not only create a body of accurate data but also create connections between stakeholders and a forum in which they can articulate what types of information are important to them. In other words, knowledge sharing may play a role in creating or strengthening trust between riparians.

Prasai and Surie argue that the exchange of basic information can help communities better understand what is happening upstream or downstream. It thus creates empathy, which is important because perception matters in politics. Conflict often intensifies when one side poorly understands the other. The proactive release of information can help to moderate the positions of both parties, argue Prasai and Surie (2013, p. 31), since finding common ground starts with having a common understanding of the situation. They show that 'local communities harbor an understanding of the ecology of the basin only in reference to the events and processes that occur within their immediate environs' (Prasai & Surie 2013, p. 31). The consequences of this localised understanding of ecology is that people downstream have no way of knowing whether a recent flood was caused by excessive rain upstream, a glacial outburst or human action at a particular barrage. This limited understanding of the basin's ecology makes it difficult for communities to generate a common platform of action or to propose solutions that have broad applicability across the region. In other words, they do not have a holistic and coherent 'counter narrative' to propose to the state for inclusion in bilateral negotiations (Prasai & Surie 2013, p. 31).

Although improved knowledge sharing is one of the aims of international actors in the Ganges–Brahmaputra problemshed, they have not (yet?) significantly altered the culture of secrecy around hydrological data.

9 The small issue of the big brother

The effect of India's hydro-hegemony

In addition to the obstacles to transboundary water cooperation discussed in Chapter 8, India's hydro-hegemony is a major challenge to improving interactions between riparian states. In large part, the distrust of India is rooted in the perception of India as a bully – a domineering hydro-hegemon, rather than a constructive leader. This chapter assesses India's hydro-hegemony and discusses the effects it has on transboundary water cooperation in the Ganges–Brahmaputra problemshed.

The first of five sections discusses the perceptions of hegemony in the region. The resentment of Nepal, Bhutan and Bangladesh towards this riparian 'bully' impedes cooperative transboundary water interactions. The second presents the theory of hydro-hegemony and the third examines India as a hydro-hegemon: India is neither a bully, nor a leader. The fourth section addresses India's preference for bilateralism (as opposed to multilateral approaches to transboundary water issues). Strictly bilateral arrangements contravene the principles of good water governance. The fifth and final section of this chapter proposes some approaches for countering India's hydro-hegemony.

Perceptions of hegemony

The hydropolitics of the Ganges–Brahmaputra problemshed are, according to Hill (2008, p. 75), defined by the dominance of India, and the question of whether transboundary disputes should be handled bilaterally or internationalised continue to provoke tensions. India's relative political, military and economic power allows it to mitigate the fact of geography that it does not control the headwaters of all the rivers that pass through its territory. India is able to influence its upstream riparians – Nepal and Bhutan – to gain access to their water resources, and its downstream riparian – Bangladesh – to overlook transboundary water arrangements that may adversely affect it. As Earle et al. (2015, pp. 41–42) show, many consider India to be the regional bully and thus the trust deficit in the Ganges–Brahmaputra problemshed is relatively high.

Nepal, Bhutan and Bangladesh are among the poorest nations on earth with an extremely limited supply of cultivable land; water will always remain a key political issue with them. Yet, as Ray (2008, p. 88) argues, India's political relations with these countries, except for Bhutan, are far from cordial. Furthermore, India is not bound to abide by the internationally accepted principles of equitable and reasonable utilisation, and an obligation not to cause appreciable harm (Biswas 2001, p. 13). In 1997 India abstained from voting for the United Nations Convention on the Law of Non-Navigational Uses of International Watercourses (UNWC) and has not since then signed on to this foundational instrument of international water law (Ray 2008, p. 83).

While some of the distrust towards India is in response to India's sheer geographical and economic size, a lot of it also comes from India playing a dominant role in the region (Singh 2008, p. 35). Indeed, as Jain (2010, p. 159) claims, all the states of South Asia, except Bhutan, look upon India with suspicion. This is because, as Bhasin (2008, p. 7) explains, India expects to hold the position of *hegemonia* (conferred headship), while the rest of South Asia view India's position as *arche* (asserted headship). Both are arguably true of India at different times and in different contexts. In any case, as Sneddon (2013, p. 16) argues, it is more useful to think of multiple hegemonies or hegemonic projects rather than a single hegemony because hegemony acts in a nuanced way and the way consent is manufactured is constantly being rearticulated. Zeitoun et al. (2014a, p. 4) add that just as conflict and cooperation co-exist, so do hegemony and counter-hegemony.

What is hydro-hegemony?

In the context of transboundary waters, Cascão and Zeitoun (2010, pp. 31–32) argue that hegemony rests on four pillars: material power (political, economic and military power, technological prowess, international political and financial support, and so on), geographical power (upstream or downstream riparian position), bargaining power (capability of actors to control the rules of the game and set agendas), and ideational power (capacity of a riparian to impose and legitimise particular ideas and narratives). The overall balance of power along these pillars determines each riparian state's degree of control over shared water resources. Moreover, Zeitoun and Warner (2006, p. 436) show that 'upstreamers use water to get more power, downstreamers use power to get more water'.

In the case of the Ganges–Brahmaputra problemshed, India is the hydro-hegemon because it has the most material, bargaining and ideational power. The strength of these pillars more than offsets the downstream position vis-à-vis Nepal and Bhutan (India is also upstream of Bangladesh). India uses this power to get more water (in the form of hydropower, dry-season flow regulation and flood control) from upstream Nepal and Bhutan. At the same time, India is able to develop its own water resources without taking into account downstream Bangladesh, which is in the overall weakest position in terms of the four pillars of hydro-hegemony. Nonetheless, as Mitra (2003, p. 399) points out, India does not use its power to overwhelm its neighbours, but instead inadvertently creates a sort of political stalemate in South Asia.

The theory of hydro-hegemony, according to Zeitoun and Warner (2006, p. 455), posits that 'the hydro-hegemon can establish the form of interaction over transboundary waters that it prefers'. It does so through coercive, utilitarian, normative or ideological mechanisms (Zeitoun et al. 2014a, pp. 3–4). The first of these mechanisms means that compliance with the order is assured through pressure created by the use of force or the threat of the use of force, for example, the unilateral construction of infrastructure. Utilitarian compliance-producing mechanisms are used to encourage compliance with the *status quo*. These are the 'carrots' to the 'stick' of coercive mechanisms. Utilitarian mechanisms can include inducements or rewards that bring either political pay-offs, such as recognition and similar rewards and alliances for legitimacy-challenged regimes, or financial pay-offs such as a share of the capital generated by a transboundary water project.

Normative compliance-producing mechanisms rely on instilling the belief that compliance with the order is right, or a duty, even an obligation. Transboundary water treaties, 'best practice' of transboundary water management, and the operational procedures of international financial institutions may be considered examples of normative compliance-producing mechanisms. Ideological hegemonic compliance-producing mechanisms rely on unquestioned acceptance of the order, independent of the sense of obligation, or rewards or pressure on offer. When these mechanisms are effective, not complying with the order is *unconsciously* ruled out (Zeitoun et al. 2014a, pp. 3–4).

Neither leader nor bully: India as hydro-hegemon

India has on the whole, according to Zeitoun and Warner (2006, p. 437), created compliance through utilitarian, normative and ideological mechanisms rather than relying on coercion to maintain the *status quo*. As such, they argue, India's hydro-hegemony has been a factor in preventing rather than causing violent conflict. But this understanding of hydro-hegemony – as a phenomenon that may lead to positive-sum outcomes as well as zero-sum ones – is often missing from water conflict analysis, claim Zeitoun and Warner (2006, p. 439).

Applied judiciously and with foresight, hydro-hegemony could translate into constructive regional leadership. The hydro-hegemon can take on a guiding, leadership role in the basin and establish positive interaction with co-riparians, for example through an integration strategy. But the hydro-hegemon can also engage in unilateral, negative, exploitative interactions, such as resource capture or containment strategies, and exert domination. Which path is chosen when depends on factors other than those related to transboundary water interaction. As Zeitoun and Warner (2006, p. 455) point out, '[w]hether the hegemon chooses leadership or domination in the water sector is ultimately governed by the broader political context'.

In the political context of the Ganges–Brahmaputra problemshed, India vacillates between being the 'regional bully' and the 'regional pushover', according to Mitra (2003, p. 401). India, claims Bhasin (2008, pp. 19–21), tends to support or endorse rather than initiate and facilitate regional ventures. It rarely champions the group interest at regional fora, instead preferring to address and manage issues at the bilateral level. If the defining feature of a hegemonic power is the ability to induce change in the policies of the target group so that the interest of the dominant nation is served, argues Bhasin, then India is not a fully realised hegemonic leader. Instead, India's policies reflect supervisory postures rather than a more fundamental participatory leadership character. Bhasin (2008, pp. 19–21) continues that India has failed to provide political advice and economic guidance to neighbours without appearing domineering.

In short, India has not (yet?) become a constructive regional leader on transboundary water issues; but it has also not realised overwhelming domination of the Ganges–Brahmaputra problemshed. Neither India's power variable nor its policies have succeeded in inducing transboundary cooperation in the Ganges–Brahmaputra problemshed. But the support, if not the leadership, of the hydro-hegemon is critical in any transboundary initiative because without it very little can get done. This is especially so in the Ganges–Brahmaputra problemshed where Nepal, Bhutan and Bangladesh do not share borders with one another and thus are limited in developing transboundary initiatives among themselves without India's participation or consent. Lack of active support from the hydro-hegemon for increasing transboundary cooperation has also undermined the work of international actors in this regard (a point discussed in Chapter 7).

India has been able to limit transboundary water cooperation in the Ganges–Brahmaputra problemshed because, as Warner and Zawahri (2012, p. 218) show, in asymmetric power structures:

> strong states have the option to select the approach consistent with their best interests, while significantly weaker states can rarely afford to opt out of cooperation or communication. Consequently, cooperation under these conditions can limit the room for manoeuvre of non-hegemonic powers.

They cite the Ganges basin as an example: India has been able to govern this international basin through strictly bilateral – rather than multilateral – accords, and this has prevented Nepal and Bangladesh from offsetting the hydro-hegemon's power. It has also prevented the riparians from developing the basin in an integrated manner recommended by scientists for the basin's efficient development (Warner & Zawahri 2012, p. 218).

India's preference for bilateralism

India's refusal to take up any river management issue in a multilateral forum, claim Earle et al. (2015, pp. 41–42), limits the possibility of a River Basin Organisation (RBO) to be established. Hill (2008, p. 60) argues that India's persistence in maintaining purely bilateral arrangements and not involving the international community in matters of transboundary water governance in South Asia has certainly been decisive in shaping the politics of water sharing in the region. He adds that bilateral treaties have been a source of considerable enmity between India, Bangladesh and Nepal (Hill 2008, p. 60).

Even though all the rivers flowing through India are international and pass more than one country, all the treaties on these rivers are bilateral (Singh 2008, p. 36). India did not even acknowledge that usage of the Ganges is an international issue until 1970 (Hill 2009, p. 91). On the other hand, Nepal, Bhutan and Bangladesh all share political borders with India, but not with each other (Chellaney 2011, p. 277). This makes it obvious and important for them to engage in treaties directly with India, but not each other. Yet tensions remain about whether transboundary water disputes should continue to be handled bilaterally (the hydro-hegemon's prerogative) or internationalised (Hill 2008, p. 75).

Tellingly, India's current National Water Policy (2012) devotes only two paragraphs to transboundary waters, and these emphasise bilateralism as the sole approach. This need not be surprising; hegemonic powers benefit through bilateral arrangements, while small and medium powers enjoy greater leverage within multilateral institutions (National Research Council of the National Academies 2012, p. 91). Yet the lack of multilateralism in South Asia regarding transboundary water management is not solely a manifestation of India's hegemonic ambitions. The more riparian states that are involved, the more difficult become the processes of multilateral diplomacy and negotiation – of finding common ground for reaching agreement on collective action, norms or rules (Ray 2008, p. 84). In the interest of avoiding conflict, bilateralism is a faster and surer approach than multilateralism. For similar reasons, it may in some cases be more appropriate to forgo a basin-wide approach to transboundary water cooperation (that is, a framework wherein all riparian states must reach consensus) in favour of a problemshed approach (that is, a framework where only stakeholders in the issues at hand must reach consensus, irrespective of whether they are within the hydrological basin or not).

Some issues that affect all the riparians may nonetheless be better managed locally, even if they do cross borders. For example, flooding and sedimentation control are problematic issues for all the riparians in the Ganges–Brahmaputra problemshed, but they do not necessarily require multilateral agreements. As Sadoff (cited in Colopy 2012, p. 320) points out: 'If you can't reduce floods in Bihar or saline intrusion in the Sundarbans with infrastructure in Nepal, then that simplifies the discussion. Maybe you don't have to negotiate everything across boundaries.' Conceptualising water issues within problemsheds may be more appropriate than within basins or political boundaries.

There are other problematic consequences of bilateral agreements. They may belie conflict and asymmetry of outcome; treaties are not the embodiment of unequivocal cooperation (Zeitoun & Mirumachi 2008, p. 300). There are usually power asymmetries at play between riparian states that can lead to less than equitable outcomes for the weaker party. The hydro-hegemon may structure a treaty to reflect existing inequalities and use utilitarian mechanisms or soft power to attain the weaker riparian's compliance (Zeitoun & Warner 2006, pp. 447–448).

An example of a hegemon enshrining inequalities in international agreement is the Treaties of Friendship that India instigated in the region. The treaties arose at a time when India gained independence from the British and wanted to prevent a communist influence spreading south from China. Consequently, India concluded what Subedi (2005, p. 42) characterises as three lopsided treaties with its small neighbours, Nepal, Bhutan and Sikkim in order to bring them into its sphere of influence in order to make the Himalayas a strong natural border with China. A similar treaty was signed with Bangladesh in 1972 (Condon et al. 2009, p. 9).

India has since been reluctant to update any of these treaties (Bhasin 2008, p. 16). But while this is a source of some discontent (especially in Nepal) (Subedi 2005, p. 5), there has been no real momentum from these smaller states to agitate for change. One tactic available to the non-hegemon is that of issue linkage wherein non-water-related issues are including in negotiations over water as a point of leverage (Zeitoun & Warner 2006, pp. 454–455). This, however, has not been a tactic utilised by India's co-riparians to alter the *status quo* of the Treaties of Friendship.

Another example of the hydro-hegemon instigating a treaty that institutionalises an inequitable *status quo* is India's signing of the double bilateral Ganges River treaties with Nepal and Bangladesh. The treaties helped maintain the *status quo* in favour of India and Zeitoun et al. (2011, p. 163) view them as exclusionary devices within an inclusive process, which did confer some benefits onto the other, weaker riparians. Condon et al. (2009, p. 40) argue that India is likely to continue its pattern of resolving water disputes with Bangladesh through bilateral negotiation rather than a regional approach. As hydro-hegemon, it is able to set the terms of cooperation and Bangladesh is likely to continue consenting to this stable arrangement.

Countering India's hydro-hegemony

India's insistence on maintaining strictly two-sided relations has not been successfully challenged by individual states, sub-regional groupings or multilateral institutions in South Asia, although opportunities exist for the development of such bulwarks to India's hydro-hegemony. Dash (2008, p. 111) identifies four ways in which states can constructively counter hegemony. The first is by creating sub-regional groupings to improve their balance of power vis-à-vis a regionally dominant or threatening state. The second is regionalist entrapment; states support the creation of regional institutions to limit the free exercise of hegemonic power. The third is called bandwaggoning, which means that weaker states seek accommodation with the regional hegemon in order to receive military and economic rewards. The final approach is declining hegemony, or forcing the hegemon to initiate common institutions to pursue its interests, to solve common problems, and to generate international support and legitimacy for its policies (Dash 2008, p. 111).

Nepal, Bhutan and Bangladesh have not created a sub-regional grouping to counter India's hydro-hegemony, and thus are not engaging in Dash's first approach. The second approach has not been successful in the Ganges–Brahmaputra problemshed either. The only regional institution that could counter-balance India's hydro-hegemonic ambitions, the South Asian Association for Regional Cooperation (SAARC), is weak and ineffectual (Dash 2008, p. 120). The fourth approach has been applied – but with limited success. Attempts at involving India in solving common water problems through multilateral cooperation – for example, by Nepal and Bangladesh over the Ganges River (Hill 2008, p. 73), or by China over the Brahmaputra (Hangzo 2012, p. 5) – have been *ad hoc*.

Dash's (2008, p. 111) third way of countering hegemony, that of bandwaggoning, is the one that has been utilised most successfully in the Ganges–Brahmaputra problemshed. Even so, it has only been used to maximum advantage and minimum anxiety in one case, that of Bhutan. The kingdom has been able to successfully seek accommodation with the hydro-hegemon to receive political, military and economic rewards. Dash (2008, p. 116) notes that Bhutan and the Maldives remain the only SAARC members that have never sought external assistance to limit India's exercise of hegemonic power in the region.

External or third parties may also be able to counter hydro-hegemony. Zeitoun and Jägerskog (2009, p. 12) write of international organisations and foreign aid donors challenging power asymmetry in two ways: they can either level the 'players', or level the 'playing field'. I return to these strategies in Chapter 12. But international actors in the Ganges–Brahmaputra problemshed seem unwilling or unable to address the issue of India's hydro-hegemony and the effects this has on transboundary water cooperation. This, and other limitations on international actors in improving transboundary water interactions are discussed in Chapter 10.

10 Limitations of foreign-led approaches

International actors working to improve transboundary water governance in the Ganges–Brahmaputra problemshed have an important – but challenging – task. Some of the obstacles to positive transboundary water interaction in this region were discussed in Chapters 7, 8 and 9. This chapter turns to the limitations inherent in the way the World Bank and other foreign aid donors operate.

For a start, creating knowledge about the Ganges and Brahmaputra basins and facilitating the sharing of it among riparians is only one small component of increasing transboundary water cooperation. Similarly, taking a basin-wide approach to transboundary water governance may be appropriate for some of the issues in this region – but not all.

A further limitation on the conduct of international organisations and foreign aid donors is that they do not adequately address some significant contextual factors that contribute to the negative or neutral transboundary water interaction between riparians. For example, there is little consideration for how low political will within the region undermines the possibility of success of these initiatives. The structural power asymmetry between India, Nepal, Bhutan and Bangladesh is another factor not generally considered by foreign-led interventions.

The nature of water conflict and cooperation is also poorly understood, and thus the prevailing theory of change is that improved transboundary water interaction will also improve regional stability. The root causes – historical grievances and perceptions of injustice – remain unaddressed by international actors. Indeed, the principles of water justice seem not to have been widely considered by third parties in the design of interventions into transboundary water governance in the Ganges–Brahmaputra problemshed.

Finally, the main method for improving transboundary water interaction, that of Track II dialogue, is a necessary but not sufficient condition of water cooperation. Such dialogue is only one factor of positive transboundary water interaction, but in the absence of other factors – such as political will and trust among riparians – is unlikely to increase water cooperation. These limitations on foreign-led interventions are discussed in turn in this chapter.

Knowledge sharing is not the solution to all transboundary water governance problems

Creating knowledge about the problemshed, and sharing it among riparian stakeholders is one of the key tasks of international actors in South Asia. It underpins good water governance and fills a much needed information gap. Yet, knowledge creation led by international actors is not significantly improving transboundary water interactions in the Ganges–Brahmaputra problemshed. As Zeitoun et al. argue, on the surface, the outputs in the form of joint reports and publications may provide some new perspectives to transboundary water resources management but do not necessarily transform the existing transboundary water arrangements. This is because these reports do not challenge the fundamental power relations between riparian states. These international efforts are thus 'taking the reformist approach that is not likely in the future to transform transboundary water arrangements or to resolve the transboundary water conflict' (Zeitoun et al. 2016, p. 14).

River basin modelling is another activity favoured by international actors. For example, Australia's e-Water initiative is bringing river basin modelling technology used in the Murray-Darling basin to part of the Ganges in India (eWater 2012). The technology itself is impressive, but it may not be the whole solution to India's river management problems. As the former head of the South Asia Water Initiative (SAWI) acknowledged, India requires more than just river basin modelling technology; it needs a whole package of modelling combined with capacity building, strengthening of governance institutions and so on (Young 2014). The Australian government has made some small pushes in that direction, and in 2015 established the Australian Water Partnership (AWP) to support public and private partnerships for sharing Australia's water sector expertise in the Indo-Pacific region (AWP 2016). But improving water management, even with the best expertise in the world, is a long, arduous and political process (Young 2014). For the time being, therefore, focusing on tangible, technological approaches is an easy win for international actors even if it does not contribute much to improved water governance in the region.

The limited hydrological data is not a major problem in the Ganges–Brahmaputra problemshed – lack of hydro-diplomacy is (Anonymous 7, 2014). River modelling alone, therefore, is not enough to improve water governance in the region. Siddiqui (2014) adds that it is especially problematic to put faith in the river basin modelling technology of foreigners who are not familiar with the peculiar hydrology of South Asia, that is, models that are not created specifically to account for the vagaries of monsoonal rivers and glacial melt. Modelling is also useless when it is based on incomplete or inaccurate government data, or when the data is unavailable – and this is too often the case in the Ganges–Brahmaputra problemshed (Siddiqui 2014).

Yet, there is some utility to river basin modelling in terms of building transboundary water cooperation. River basin technology should not be seen just as the output of data but rather as the process of developing the model. Bringing together scientists, engineers and community stakeholders to discuss what is important to them in the basin and their assumptions about cause and effect within the system is critical to developing meaningful cooperation and local buy-in. This is understood by some international actors intervening in the hydropolitics of the Ganges–Brahmaputra problemshed.

The Australian Commonwealth Science and Industry Research Organisation (CSIRO) is currently funding a project on the science–policy interface in the Kosi basin, a sub-basin in the Ganges–Brahmaputra problemshed. The project is run by the International Centre for Integrated Mountain Development (ICIMOD) and is based on the assumption that sound hydrological modelling is important in establishing Integrated Water Resources Management (IWRM). Various water policy practitioners, academics, scientists, hydrologists and so on are consulted during the project about their water governance priorities (Siddiqui 2014). This is an example of how the process of developing river basin models may indeed facilitate dialogue and cooperation between stakeholder groups who may otherwise have not engaged with one another on issues of mutual importance.

Nonetheless, there are many misconceptions and cultural barriers to overcome before the process of river basin modelling can build transboundary water cooperation. For a start, as Siddiqui points out, community involvement is considered a nuisance by many in the scientific community; hydro engineers, for example, may decide the best place for an irrigation canal based on scientific measurements but the local community may object for reasons of local politics or traditions (Siddiqui 2014). These cultural barriers speak to the need for improved hydro-diplomacy at all levels and illustrate that dialogue is a prerequisite for the success of more technical solutions to transboundary water issues.

A further barrier is that of secrecy around hydrological data that does exist. As discussed in Chapter 8, transboundary water cooperation in the Ganges–Brahmaputra problemshed has often been hindered by the limited transparency around hydrological data; where information about shared rivers does exist, it is not readily shared. Significantly, Prasai and Surie (2014) point out that the reluctance to share such data is not a matter of policy but rather of culture. This culture of secrecy is what international actors are attempting to change in order to increase cooperation of the region's transboundary waters.

SAWI, for example, is trying to bridge the gap in data sharing, and is succeeding in doing so at the people-to-people level (though progress on the government-to-government level is proving more difficult) (Anonymous 5, 2014). The World Bank as a whole has for the past decade styled itself as an 'information bank'. It produces much of what Roy (2014) calls 'well written rubbish' that is sometimes of questionable methodological soundness but nevertheless brings important information into the public domain. ICIMOD too has been instrumental in breaking down the culture of secrecy around hydrological data by conducting, for example, meteorological data sharing projects between Nepal, Bangladesh, Bhutan and Pakistan (Molden 2014).

These foreign-led interventions on the whole produce good data and interesting reports that raise awareness of important issues (such as the impact of Glacial Outburst floods (GLOFs)) and bridge knowledge gaps. But the pace of progress is not as per expectations. One Nepalese stakeholder argues that perhaps international organisations and foreign aid donors should be instead focusing their efforts on Track I processes for data sharing, that is, government-to-government interactions (including state government to state government in India). This would likely speed up the improvement of relations between countries in the Ganges–Brahmaputra problemshed (Anonymous 5, 2014) and contribute significantly to the breakdown of the culture of secrecy. As it is, however, the Track II dialogue and low-level scientific data sharing initiatives are having only a limited effect on overall transboundary water interaction.

Improving how information is shared between governments may also have only a limited effect on improving transboundary water interaction. As D'Souza (2014) points out, hydrology is only one facet of rivers, and relationships are more important than data. Furthermore, what a river means to a fishing community is different to what it means to a farming community, and thus their information needs will be different (D'Souza 2014). A focus on improved communication between stakeholders with complementary information needs may be a more effective tactic than improving official channels. After all, as Prasai and Surie (2015, p. 28) have shown, Freedom of Information Acts do exist, but that has not necessarily resulted in improved communication or information flow between those who need it most.

Taking a basin-wide approach is not always appropriate

Part of the focus on generating and sharing data about the Ganges–Brahmaputra problemshed is part of the philosophy of Integrated Water Resources Management (IWRM), which stresses the importance of a basin-wide approach to water governance. This approach has some fans within the region. Sharma (2014) sees a basin-scale approach as appropriate for the rejuvenation of the Ganges – a massive undertaking that has gained momentum in recent years.

A basin-wide approach is part of the underlying philosophy of foreign-led intervention. The Abu Dhabi Dialogue (ADD) brought together all riparian states from the Indus, Ganges and Brahmaputra basins, and so does SAWI. Earle et al. (2015, p. 41) argue that basin-based institutions could help to reduce uncertainty surrounding Ganges–Brahmaputra riparian relations. The theory of change is that such institutions may be able to shift the focus from disconnected and short-term interactions of the riparian countries into a continuous relationship that has scope for future routine gains.

The theory, then, is sound: basin-wide approaches can build trust and provide a forum for the resolution of transboundary water conflicts. But in the decade of the Abu Dhabi Dialogue this has not eventuated. In reference to the Ganges–Brahmaputra–Meghna basin, Earle et al. (2015, p. 41) point out that these expectations have not resulted in much beyond political statements and 'setting up lofty goals'.

A basin-wide approach, furthermore, may not be an appropriate solution for many other problems in the Ganges–Brahmaputra–Meghna basin. Swain (2011, p. 696) shows that a basin-wide approach was not successful in the Nile Basin Initiative, on which the Abu Dhabi Dialogue was modelled, and that a sub-basin approach may have had more of a chance of successfully increasing transboundary water cooperation in the Nile basin (Swain 2002, pp. 293–308). The basin-wide approach requires consensus from all riparian states, and that can be an impediment for taking action on issues that do not affect the basin as a whole, for example flood management. For this reason considering the Ganges–Brahmaputra problemshed as a separate and distinct unit of water management from the Ganges–Brahmaputra–Meghna basin may be a more functional scoping of transboundary water governance issues.

Benefit sharing may also be limited by the complications of balancing benefits across an entire basin. As Sadoff and Grey (2005, p. 422) point out, '[i]f significant benefits accrue in one country, while significant costs are borne by another, it is possible that a project providing net benefits on a basin-wide scale could actually generate net losses in any one country'. This, in turn, may compound perceptions of water injustice and thus perpetuate disputes between riparians rather than contributing to cooperation between them. Significantly, Zeitoun and Mirumachi (2008, p. 310) list benefit-sharing based on agreements as an example of neutral rather than of positive interaction. In other words, benefit sharing is not a straightforward solution to increasing cooperation between riparian states.

Political realities of the region are not accounted for

International actors have found an opportunity to intervene in the transboundary water governance in the Ganges–Brahmaputra problemshed because riparian states seem unwilling or unable to address shared water issues. But without political will for change, it is difficult if not impossible to alter the *status quo*, especially from outside the system. Earle et al. (2015, pp. 41–42) point out that there is certainly a lack of political will in the basin, particularly in India. Another regional water analyst agrees and adds because there is little political appetite for transboundary cooperation within these countries (among their ruling elite), and even less in India, the work of international actors in this regard becomes more challenging (Anonymous 6, 2014). In other words, the drive for transboundary cooperation comes almost exclusively from international actors, rather than from the riparians (Anonymous 6, 2014). The historical, cultural and political reasons for this are not addressed.

As Hill (2013, p. 257) argues, multi-scalar civil society networks could play an active and important role in overcoming the impasse of lack of political will. But at present such networks are weak, underdeveloped and unable to exert significant influence on state policy and inter-state relations. Moreover, foreign-led interventions are not explicitly building these sort of networks, but rather focusing on the limited set of relationships within and between government actors through Track I and II dialogues. While that is one part of the puzzle of transboundary water cooperation, the efficacy of SAWI and the Abu Dhabi Dialogue is weakened by India's low commitment to multilateralism (Thakkar 2014).

A further limit on the efficacy of international actors is that they are divorced from the political reality in the region. As one strategic analyst explained, international actors must operate with the dominant political context and not simply assume a good idea will come to pass because it is a good idea; it must be politic too. These third parties must take into consideration all the 'political horse trading' that goes on and which can either facilitate or hinder their initiatives. According to this same analyst, SAWI does not seem to operate astutely within the political context of the Ganges–Brahmaputra problemshed (Anonymous 4, 2014). Yet, without understanding the current as well as historical political context, half a century of ill-will relating to shared waters cannot be undone, argues Prasai (2014).

Tiwary (2006, pp. 1686–1687) shows that considerations outside the negotiation table affect the outcome of formal negotiations. He argues that above all else, international water conflict resolution requires considerable political will on all sides and sustained motivation to reach agreement. Tiwary (2006, pp. 1686–1687) also argues that states negotiate not only in reaction to other riparians, but also in reaction to domestic politics. For example, in countries where threats of water security are more apparent, or in a relatively weaker state that strikes a deal with stronger states, there is always the risk of negotiators being labelled as villains who compromised over national well-being (Tiwary 2006, pp. 1686–1687). The fear of this outcome colours the negotiation dynamic. Thus, understanding the cultural context and power dynamics between riparians is important in understanding informal negotiations over transboundary waters.

A significant part of the cultural and political context in the Ganges–Brahmaputra problemshed is the presence of a hydro-hegemon. Yet, the structural power asymmetry between India, Nepal, Bhutan and Bangladesh is not factored into the Track II dialogue processes being facilitated, and there is little consideration for how lack of political will within the region undermines the possibility of success of these initiatives.

As discussed in Chapter 9, hydro-hegemony and power asymmetry between actors can be a factor of both conflict and stabilisation. Yet, the international actors engaged in increasing regional stability through cooperative transboundary water governance do not incorporate hydro-hegemony theory into their theory of change, nor actively address the effects of the power imbalance between India and its Ganges–Brahmaputra problemshed co-riparians. As the hydro-hegemon in the region, India is able to set the agenda for water governance and this inhibits its relations with neighbouring states, who resent and brace against India's dominance (Prasai 2014).

The presence of external, foreign actors sometimes provides a much needed neutral party to water negotiations (Novel 2014), but overall the international initiatives are unlikely to significantly alter the power dynamic or perception of inequality in the Ganges–Brahmaputra problemshed. Moreover, the hydro-hegemon's support for any interventions into the hydropolitics of the region is crucial for their success. Yet, as discussed above, India has been only a lukewarm supporter of the Abu Dhabi Dialogue. India's hydro-hegemony rather than the power of consensus continues to set the tone for the hydropolitics in the Ganges–Brahmaputra problemshed. As Wirsing et al. (2013, p. 13) point out, 'far from heralding a new and more cooperative era, the situation seems reflective of an all too familiar state-centric and asymmetrical power-political order' and does not bode well for the region's future water security.

Understanding the dynamics of hydro-hegemony – and especially counter-hegemony – can provide the basis for a strategy to alter the *status quo* (Zeitoun et al. 2014b, pp. 187–188). Power asymmetry, according to Zeitoun et al. (2014b, p. 174), enables or disables justice claims and conflict resolution efforts. Hegemonic actors use their power to 'deny or curtail the claims by non-hegemonic actors of past injustices, or of current inequitable arrangements of water use, treaties, or institutions' and thus conflict resolution is most effective if based on analysis that appreciates and brings to life these intricacies of justice (Zeitoun et al. 2014b, p. 175). But these intricacies of water justice and the effect of power asymmetry on conflict is not expressly taken into account by SAWI, or the two main international donors of transboundary water governance, Australia and the United Kingdom. How a water justice approach can be incorporated into foreign-led initiatives is discussed below and in Chapter 12.

Increasing water cooperation will not necessarily result in greater regional stability

International organisations and foreign aid donors engaging in the Ganges–Brahmaputra problemshed are not informed by the literature on natural resource conflicts and water interactions. The foundational documents of SAWI, and the British, Australian and Norwegian initiatives in the Ganges–Brahmaputra problemshed do not reference even the most prominent scholars in this field, such as Aaron Wolf or the London Water Research Group's work on hydro-hegemony. Some documents by other research/policy organisations, such as the Asia Foundation (see for example Prasai & Surie 2013) or Adelphi (see for example Pohl et al. 2014), are informed by this research, but these are not prevalent or significant enough to meaningfully influence discourse about water conflicts and transboundary water governance in the Ganges–Brahmaputra problemshed. This is important because, as Keohane and Nye (2001, p. 7) warn, applying the wrong rhetoric to problems leads to erroneous analysis and bad policy. It also undermines the intent of deliberative governance, and Track II dialogue, which theoretically ought to include academics.

There have been several swings in the thinking on water conflict and cooperation. The work of Aaron Wolf in the 1990s has been instrumental in refuting the once prevalent water wars thesis. However, this has led to an arguably over-optimistic mood that sees cooperation as the inevitable march of history. More recently, scholarship such as Mirumachi and Allan's Transboundary Water Interaction Nexus (TWINS) model, and Zeitoun and Warner's (2012, p. 217) work on hydro-hegemony have served as cautionary tales against the prevalent optimism without reverting to the water wars scenario.

Zeitoun and Mirumachi (2008, p. 32) also offer an important point for the application of deliberative water governance initiatives and the evaluation of their success in contributing to regional stability: cooperation and conflict are not binary opposites. The reliance on continua for plotting cooperation versus conflict in a linear dichotomy has unintentionally led to the prevailing understanding that all conflict is 'bad' while cooperation is inherently 'good'. Significantly, this tendency has led to policy assuming that relations over transboundary waters may be improved simply by promoting movement along one direction or on one element of the water interaction, rather than addressing the broader context as a whole (Zeitoun & Mirumachi 2008, p. 32). This is the case in the Ganges–Brahmaputra problemshed.

Just as the nature of cooperation is poorly understood by international actors, so the nature of conflict is misunderstood. In South Asia, as elsewhere, the absence of war is not equivalent to an efficient and equitable solution to water governance problems (Singh 2008, p. 32). A sense of unfairness and injustice is often at the root of conflict, over water or otherwise (Upreti 2004, p. 60). For example, tensions may develop between those who are rich enough to cope with water shortages, and those who are too poor (Cosgrove 2003, p. 3). As such, the goal of creating equitable solutions and establishing water justice may be a more appropriate one for foreign aid donors and international organisations aiming to increase stability and cooperation in the Ganges–Brahmaputra problemshed.

The assumption underpinning the internationally driven initiatives for deliberative river governance in the Ganges–Brahmaputra problemshed is that if countries cooperate on common water issues they may be more likely to cooperate on other issues. Cooperative transboundary water interactions are assumed to translate directly into overall cooperative geopolitics. This assumption comes from functionalist thinking within international relations, which posits that cooperation on transnational issues, such as trade and environment, will spill over to other more difficult issues. As Dinar et al. (2007, p. 146) argue, 'cooperation on water issues could not only be easier to achieve than other more political and protracted issues, but also may form the basis for trust and cooperation on more difficult issues'.

This theory, however, is not borne out empirically. India and Pakistan, for example, have a formalised and relatively cooperative (albeit highly politicised) relationship over the rivers of the Indus basin under the auspices of the Indus Waters Treaty, which has withstood wars between the two countries that were not directly related to water (Hill 2009, p. 89). Yet, the overall relationship between India and Pakistan is very tense. Conversely, countries that have conflicted hydropolitics – such as those in the Ganges–Brahmaputra problemshed – may well have more peaceful relations in other spheres. In other words, violent conflicts over non-water related issues can emerge between states even as water interactions remain cordial and *vice versa*.

Another problematic aspect of the theory of change for regional stability through improved water governance is the assumption of a spill-over effect from cooperation over rivers to cooperation over other disputed sectors. Creating an understanding of benefit sharing, positive-sum outcomes and so on in transboundary water governance is expected to translate into an overall management of natural resources as a common good, and thus reduce disputes over other contested resources. That, however, is a paradigm shift that is impossible to achieve with any one project or initiative; it requires slow, incremental and messy cultural change and genuine, consistent political will from within the region and commitment from the ruling elite to these principles.

Moreover, increasing transboundary water cooperation may mitigate certain types of conflict and insecurity while exacerbating others. Social and economic interdependence between states, for example, has for a long time been considered a factor in decreasing rates of international conflict. Yet conflict does not disappear when interdependence prevails; on the contrary, conflict will take new forms, and may even increase, but the traditional approaches to understanding conflict will not explain interdependence conflict particularly well (Keohance & Nye 2001, p. 7).

Water-related conflicts may also take new, unexpected forms. For example, transboundary water sharing agreements (particularly those based on a set volume of water rather than a percentage of what is available) may become sources of conflict rather than cooperation as river flows change. This is a particular risk in regions such as the Ganges–Brahmaputra problemshed where glacial melting with the risk of above-average flows may be followed by droughts (Evans 2010, pp. 13–14).

One instance of water sharing agreements being undermined by changing river flows was discussed in Chapter 3. The Ganges Treaty between India and Bangladesh allocated fixed flow amounts at the Farakka Barrage near the border of the two countries, but quickly became a point of contention as water availability declined due to unprecedented withdrawals upstream of the barrage (Hanasz 2014b). International actors are not, however, actively engaged in revising the terms of this treaty and resolving the dispute.

A further limitation on international organisations and aid donors working to build regional stability through transboundary water cooperation is that they focus on inter-state disputes rather than sub-state ones. Yet, despite the water wars rhetoric, most incidence of violence surrounding water is at the sub-national level and can occur between sectors (such as the case with the farmers' protests surrounding the transfer of water to non-agricultural use from the Hirakud reservoir in the Indian state of Orissa) as well as between different sub-state agencies (Routray 2010, p. 36). That is not to say that the World Bank and other foreign aid donors should become involved in sub-state and local disputes in the Ganges–Brahmaputra problemshed. Rather, the salient point is that these disputes can spill over state borders (Chellaney 2013, p. 55) and therefore endanger regional stability.

In larger part, however, it is because, as Iyer (2010, p. 66) has argued, 'we are excessively preoccupied with the details of particular disputes, and fail to see river-water disputes as a subset of the larger set of water-related disputes in general, and to ask ourselves what the root causes of such conflicts are'. The root causes of water conflict lie in the real or perceived injustice of water allocation.

The root causes of water-related conflict are not addressed by international actors

Zeitoun et al. (2014b, p. 188) argue that transboundary water justice is influenced by the same dynamics that impact transboundary water conflict and cooperation, namely: the disabling or enabling role of power asymmetry; the construction of scale to promote agendas; and the use of discourse to frame trade-offs and legitimacy.

If water conflict is understood as having roots primarily in unjust water governance, then it follows that conflict cannot be resolved, and cooperation instilled, without addressing the source of that real or perceived injustice. To this end, Zeitoun et al. (2014b, pp. 187–188) draw out four lessons for taking a water justice approach to transboundary water governance:

1. Conflict resolution/transformation is more important than conflict management;
2. Equitability is a pre-condition for 'positive' cooperation and a sustainable outcome;
3. There are limits to the utility of 'participation' in transboundary water negotiations;
4. Sustainable transboundary water arrangements can be informed by considering the alternatives generated by counter-hegemonic movements, or justice analysis, because they can provide the basis for a strategy to alter the *status quo*.

Broadly speaking, these lessons have not been taken into account by the Abu Dhabi Dialogue and other international actors in the Ganges–Brahmaputra problemshed. These initiatives focus on conflict management rather than resolution, do not prioritise the principle of equitability in governance outcomes, and instead over-emphasise procedural fairness in the form of broad participation. They are also not informed by or responsive to the hegemonic and counter-hegemonic dynamics in the region.

The power of a hydro-hegemon may facilitate and expedite water sharing arrangements, but this has no bearing on how equitable these arrangements are (indeed, they are likely to be perceived as unjust) and thus does not support conflict resolution. As Zeitoun et al. (2014b, p. 188) point out, 'in cases where institutions and structures supporting hegemonic claims remain entrenched, the original unjust hegemonic arrangements would very likely be replaced by new hegemonic configurations, very possibly equally unjust'. Thus, by seeking the participation, consent and support of India, international organisations and foreign aid donors may be implicitly (albeit inadvertently) continuing the hegemonic *status quo* that is at the root of the water disputes they are also aiming to rectify. Moreover, the uncritical or unprincipled hydro-diplomacy efforts that follow the tone established by the hegemon may, according to Zeitoun et al. (2013, p. 11), dismiss a water justice approach as impractical. This is illustrated in the debacle over the World Bank's Ganges Strategic Basin Assessment. This comprehensive report was rejected by Nepal and Bangladesh in part because of a perception that the solutions proposed therein favoured the hydro-hegemon, India (Pun 2013, p. 12).

It can be seen that the international actors intervening in the Ganges–Brahmaputra problemshed have not taken a water justice approach to transboundary water governance. Their efforts largely focus on conflict management rather than resolution and do not prioritise the development of equitable water sharing arrangements. The focus on participation is laudable but of limited utility in addressing the roots of water conflict, especially as those roots lie in the vast power asymmetries between riparians. The dynamics of hegemony and counter-hegemony are not taken into consideration by the third parties intervening in the transboundary water interaction in the Ganges–Brahmaputra problemshed. Lastly, historical grievances, such as those described in Chapters 2 and 3, are not addressed by SAWI or the other third parties. Yet, as Perlman-Petersen and Wolf (2015, p. 4) argue, '[i]t is necessary both to address past and present grievances as a prerequisite for market-driven solutions'.

In summary, it is unlikely that these internationally led interventions will succeed in water dispute resolution – and thus the increase of transboundary water cooperation – without addressing these power asymmetries and taking a water justice approach.

There are limits to the utility of Track II dialogue, procedural justice and participatory decision-making

One facet of a water justice approach is that of participation. This means participation in water governance processes, and in the decision-making about water allocation. As Patrick (2014, p. 70) explains, '[w]ho decides what is just and for whom depends on who or what is excluded or included in the water-allocation decision-making process'. This is important for achieving water justice, she argues, because if social inclusion is achieved, a wide variety of perspectives on water distribution and water use will be voiced. On the other hand, argues Patrick (2014, p. 70), 'the social exclusion of any particular stakeholder group can result in injustices simply because certain opinions, needs or world-views are excluded'.

The merit of the participatory approach is understood and appreciated by international actors in the Ganges–Brahmaputra problemshed. There is a growing understanding that the water resources of the Hindu Kush Himalayan region are shared resources and must be managed as such (Vaidya & Sharma 2014, p. iv). Asthana and Shukla (2014, p. 274), for example, make the case that the governments of South Asia need to move forward with a cooperative and participatory approach to river basins and the sharing of freshwater resources. They argue that 'there needs to be a regional awareness that rivers can be better harnessed through collective efforts and recognition that cooperation is essential to alleviate threats to water security' (Asthana & Shukla 2014, p. 274). Colopy (2012, p. 319) too argues that participatory, regional cooperation is the foundation of intelligent river basin management. Hill (2013, p. 257) adds that this approach is essential if the multiple water crises in the Ganges–Brahmaputra problemshed (described in Chapters 2 and 3) are to be overcome.

Nonetheless, procedural justice is not the same as equitability in Zeitoun et al.'s (2014b, p. 184) understanding of the term. They argue that recognition and participation of all relevant stakeholders is indeed important, but that distributive justice is also a necessary condition for resolving water disputes. Furthermore, if we accept Zeitoun and Mirumachi's (2008) understanding of water conflict and cooperation as coexisting, and their argument that not all cooperation is 'good', then it follows that 'participatory processes can be just as strategic, manipulative, and coercive as these forms of cooperation ... and may be very distant indeed from any world view of justice' (Zeitoun et al. 2014b, p. 187). An example of this from the Ganges–Brahmaputra problemshed is SAWI over-stating the breadth and depth of inclusiveness in the Track II dialogues it facilitates.

In short, there are limits to the utility of participation in transboundary water negotiations. That is Zeitoun et al.'s (2014b, pp. 187–188) third principle for a water justice approach to transboundary water conflict negotiations. It is significant in the context of international organisations and donors in the Ganges–Brahmaputra problemshed because these interventions favour participation (in the form of Track II dialogue) but do so at the cost of distributive justice. As one water justice advocate argues, SAWI will not succeed in increasing transboundary water cooperation as long as it remains only a 'talk shop' (Thakkar 2014). Another regional analyst points out that while Track II dialogue is important in bringing previously disempowered groups to the negotiation table, it is unlikely to make material difference unless the interests of those groups are made to be the interests of the ruling elite (Anonymous 4, 2014).

There are other weaknesses in the pursuit of transboundary water cooperation through Track II dialogue. For instance, it may be ineffective in the short term (i.e., up to ten years). To recap, Track II dialogue is used herein to denote:

> unofficial dialogues, generally between two antagonistic parties, and often facilitated by an impartial Third Party and involving individuals with some close connections to their respective official communities, focused on cooperative efforts to explore new ways to resolve differences over, or discuss new approaches to, policy-relevant issues.
>
> (Jones 2015, p. 24)

This is deceptively simple. Resolving difference and developing new policies takes, above all else, time. Yet, as Chapter 11 points out, international actors tend not to engage consistently in such projects over the long term.

How and when these dialogue processes should be pursued is also contested. Track II dialogue can also be understood as the informal, non-binding, non-attributive discussions with government and non-state actors, and as such is the nexus between Track I dialogue (formal, government-to-government relations) and Track III (informal discussions between non-state actors only such as NGOs, academia etc.). These neat definitions, however, are not easy to apply in practice; everyone understands them differently (Young 2014).

Some scholars see Track II dialogue as the foundation of cooperative transboundary interaction, and argue that it is critical that it happens before Track I processes begin because government negotiators and diplomats in the Ganges–Brahmaputra problemshed tend to have a zero-sum approach to negotiations. Therefore, there is first a need for talks between peers who can scope the problems and begin addressing it from the perspective of shared interests (Gupta 2014). The logic is that engaging with non-government stakeholders early on in a negotiation process ensures that they can establish an agenda rather than have it imposed by external or government actors (Anonymous 1, 2014). Others see less value in Track II dialogue alone and instead believe that what drives change is collaboration between non-government groups, that is, Track III dialogue (Prakash 2014). This, however, is difficult to facilitate by third parties because it is in essence a bottom-up and organic process.

By focusing primarily on Track II dialogue, international actors are limited in the kinds of discussion – and therefore solution – they can generate. SAWI in particular engages only stakeholders who already engage with government actors. As Prasai (2014) argues, because the transboundary water governance impasse in the Ganges–Brahmaputra problemshed is between governments, not between non-state actors, there is no fresh perspective on the problem set; Track II dialogue allows only for discussion between actors who are already part of the discussion about transboundary water issues.

If riparian stakeholders are already in discussions about the governance of shared rivers, is there any need to involve international actors? Most of the literature on Track II dialogue takes the view that a third party is necessary, and Jones argues that the heart of Track II dialogue is the inability or unwillingness of the protagonists to the conflict to come together of their own accord. Furthermore, he argues, brainstorming by the parties themselves may not be enough, and under the wrong circumstances it may do more harm than good. Jones (2015, p. 88) shows that the third party guides conversations down productive paths. In essence, then, Track II dialogue is not a brainstorming session between conflicted parties but rather a specific kind of facilitated discussion:

> Absent this role, the conversation can degenerate into a frustrating bargaining session which holds little promise, especially in a highly charged,

> emotional situation where the conflict is deemed to be intractable; in such circumstances, what is there to bargain over?
>
> (Jones 2015, p. 88)

Jones' point is at the crux of this argument: if there is trust between riparians and political will to take action, then there is no need for facilitated Track II dialogue (though third parties may provide technical assistance).

There is, nonetheless, a role – albeit a circumscribed one – for international actors in this sector. They are often seen as a neutral third party, and can therefore facilitate dialogue between in-basin actors who may otherwise be distrustful of each other. Thus, while it may seem counter-intuitive that a bottom-up process is being led from above and by external actors, this may in practice have the positive effect of bringing neutrality to a situation fraught with mutual suspicion and a history of poor communication (Mittra 2014).

Donors and international organisations may also contribute to the overall slow and uncertain process of improving transboundary water interaction. Creating an understanding of water as a shared resource is no straightforward task. Indeed, it seems unlikely that a paradigm shift of that kind can be imposed at all. Progress toward greater transboundary cooperation proceeds slowly but steadily. Many organisations and actors need to work in concert, 'chipping away' (in Prasai's (2014) words) at the problem. But this slow, steady and largely uncoordinated and unplanned work is not reflected by donor language, which expects to bring about an attitude shift in one funding cycle (Prasai 2014). Jones (2015, p. 140) argues that 'real peace is made as a result of a complex and interlocking web of factors, and that Track Two, however important in getting a dialogue going, is but one of these'.

11 Structural challenges in the global system of foreign aid

This chapter turns to the system of international foreign aid and addresses some of the systemic challenges that are not specific to transboundary water governance or the Ganges–Brahmaputra problemshed but that may nonetheless be an impediment to international actors achieving their stated objectives. It examines some of the structural issues of foreign aid that affect the way that donors and international organisations fund transboundary water governance programmes, and how these programmes are designed. These issues include the cumbersome and risk-averse bureaucracies that drive such programmes; the difficulty of separating out cause from effect; the focus on short-term goals, and the problem of sustaining early successes into long-term progress; and the foreign agendas that drive international development. Each of these points is discussed in turn.

Understanding the effectiveness of foreign aid generally is important because the effectiveness of international actors in increasing transboundary water cooperation supports the justification for future interventions into transboundary water governance. In other words, if the international actors in the Ganges–Brahmaputra problemshed are unlikely to substantially or directly increase transboundary water cooperation between India–Nepal and India–Bangladesh, then the rationale for similar interventions in other problemsheds in the future is weakened.

Crowded, inefficient and risk-averse – the global foreign aid system

International actors intervening in the transboundary water governance of the Ganges–Brahmaputra problemshed are part of a complex global system of foreign development aid. Although it is not the purpose of this book to critique international aid, it is worth noting that the systemic problems with global aid are large and growing (Riddell 2014, p. ii). Some of these problems affect the way foreign donors conduct water governance initiatives. For example, Evans et al. (2015, p. 201) point out that the World Bank – one of the largest donors of aid in South Asia – is known for its 'heavy and often risk-averse bureaucratic processes – geared more toward large, loan-financed operations than toward partnership-based and grant-financed programs – which tend to impede flexibility, innovation, and speed, and therefore limit impact'.

It is not only the World Bank that is cumbersome and inefficient; the system is largely like this. Foreign aid is, according to Riddell (2014, p. 27), 'characterised by an ever-increasing number of donors overseeing a growing number of discrete projects, creating an ever-more complex web of transactions and parallel management systems, many replicating and duplicating each other' and therefore 'the overall allocation of all aid is the (unplanned and unpredictable) outcome of the cumulative voluntary decisions made by all individual donors' (Riddell 2014, p. 26). As such, '[i]nefficiencies in terms of impact and effectiveness are thus "built into" the current way that aid is allocated' (Riddell 2014, p. 26). This is true of the increasing number of international actors vying for influence in the transboundary water governance of the Ganges–Brahmaputra problemshed. There are three main donors to SAWI (Australia, United Kingdom and Norway), but these donors also each run their own water governance programmes, policy initiatives and commercial ventures (for example, Australia's eWater).

The twin problems of effect and attribution

The crowded and complex system of foreign aid means that it is almost impossible to separate out the effects of individual donors on common objectives (Riddell 2014, p. 26). This is especially relevant to general, high level, long term objectives such as increasing transboundary water cooperation. Mayne (1999, pp. 2–3) notes that the question of whether or not desired effects are occurring, and the related question of attribution (that is, who or what is responsible for bringing those effects about) is well known in both international development and public policy. Programme evaluation through social science research methods is usually the best way to answer these questions of effect and attribution, but they are time consuming and expensive (Mayne 1999, p. 3). Mayne argues that even with a thorough evaluation study it may not be possible to get accurate results about a programme's effect and attribution of positive outcomes. The purpose of an evaluation, therefore, is more about knowledge creation (that is, getting an approximate sense of whether something may be working or not) than scientific measurement (Mayne 1999, p. 3).

Even with the most robust of evaluations, there are likely to be multiple causes of and explanations for specific social effects. Mayne (1999, p. 13) explains that part of the attribution problem is that there are often alternative plausible explanations for certain outcomes – the aid programme being evaluated is just one of these. He argues that 'those who are sceptical that it really was the program's contribution that counted will point to other reasons for the observed outcome – for example, other related government programs, economic or social trends, behaviour unaffected by the program' (Mayne 1999, p. 13). In the case of the Ganges–Brahmaputra problemshed there are some factors outside initiatives led by international organisations and aid donors that may affect the quality of transboundary water interaction in the near future. Indian Prime Minister Narendra Modi, though not an unequivocal champion of multilateralism, has nonetheless been reinvigorating water agreement negotiations with co-riparians that had stagnated in recent years (the Pancheswar high dam project with Nepal and the Teesta River treaty with Bangladesh are two prime examples).

At the same time, and equally importantly, it is just as difficult to attribute failure as it is to attribute success. As Pahl-Wostl (2015, p. 32) argues, it is also unclear who is to be made accountable for failed governance notions. In increasingly diffuse and complex governance networks, the roles of governing and becoming governed become blurred. She is speaking specifically about water governance, and her point applies to international actors in the transboundary water interaction in the Ganges–Brahmaputra problemshed; just as it is difficult to attribute success to these initiatives, it is also difficult to attribute failure.

Another challenge for evaluating the effectiveness of aid programmes is that 'there is likely to be a significant delay before the intended outcomes occur and the attribution picture portrayed through tracking performance over time will not be as evident' (Mayne 1999, p. 13). Nonetheless, argues Mayne (1999, p. 13), outcomes need to be tracked over the long term, even though attribution becomes even more of a challenge on longer timescales.

Short-term effectiveness, long-term failure

Ascertaining effectiveness of individual donors on long-term goals, however, may be a moot point; a further structural problem of the foreign aid system is that sustaining immediate benefits of aid into the long term often proves challenging (Riddell 2014, p. 2). Riddell (2014, pp. 8, 12) argues that foreign aid works best for accomplishing short-term effects, but warns that 'immediate project success ... does not necessarily mean permanent success. The evidence shows that the numbers of positive project assessments tend to drop markedly when projects are examined over time: there are clearly problems in sustaining success.' Thus the very nature of the objective of improving transboundary water governance in the Ganges–Brahmaputra problemshed limits the possibility of this objective being brought about by any one initiative, or sustained into the long term.

Paradoxically, the focus of many foreign-led initiatives on quick results actually undermines the potential of aid to bring about transformational changes. This is an observation made by Andrew Natsios, former head of the world's largest bilateral agency, the United States Agency for International Development (USAID) (Riddell 2014, p. 36). The transformational approach to aid-giving means assisting a country to build and promote its own development path. But, argues Riddell (2014, pp. 45–46), the systematic priority given to achieving short-term, tangible and measurable results means that less aid is available to support more complex initiatives, the intended results of which take longer to achieve and the outcomes of which are more difficult to predict.

Even with the essentially transformational objective of improving transboundary water interaction in the Ganges–Brahmaputra problemshed, international organisations and donor-led initiatives are justified and financed according to progress on short-term, quantifiable measures that have only a superficial relation to the ultimate goal. For example, Australia's South Asia Regional Development

Program Strategy is explicit about delivering 'results' in line with major international commitments to aid effectiveness (AusAID 2013, pp. 6–7).

The Paris Declaration for Aid Effectiveness commits to 'results-oriented reporting' (OECD 2005, paragraphs 44–55), while one of the three main themes of the Accra Agenda for Action is 'Delivering results that will have real and measurable impact on development' (OECD 2008, paragraph 23). As for the Busan Partnership for Effective Development Cooperation, one of its four principles is '[a] focus on results: Having a sustainable impact should be the driving force behind investments and efforts in development policy making' (OECD 2012). Effectiveness and accountability are, of course, highly desirable in foreign aid programmes. But, as discussed above, the focus on tangible results can have the unintended consequence of driving donors to focus on short-term, tangible and quantifiable programmes. In turn this may have the effect, identified by Riddell (2014, pp. 45–46), of less aid fund being available to support complex, long-term initiatives.

A further unintended consequence of a results-orientated approach to international development is that 'value for money' must be demonstrated. This, in turn, creates a misleading relationship between funding and outcomes. The assumption becomes that the more money is thrown at a problem, the greater or surer the outcome is. A trace of this assumption can be found in the documentation about the South Asia Water Initiative (SAWI). The idea seems to be that funding for Track II dialogue workshops, conferences, meetings and so on will have a proportional effect on levels of cooperation between the riparians of the Ganges–Brahmaputra problemshed. This does not take into consideration the qualitative factors of transboundary water interaction.

In the Ganges–Brahmaputra problemshed there has been more than 60 years of geopolitical impasse on transboundary water issues. Prasai (2014) doubts that money alone can resolve the underlying tensions, distrust and wicked problems of policy. This assumption is another reflection of how far removed from local political realities international actors are at times. As discussed above, these initiatives are unlikely to succeed if they are not connected to and work within the existing political context; instead of focusing on how many workshops they can fund, donors and other international organisations would be more effective, according to one regional analyst, through understanding all the 'political horse trading' that can both facilitate and hinder initiatives for greater cooperation (Anonymous 4, 2014).

On the other hand, it can be said that SAWI is not interested in meddling in regional politics but rather in capacity building within the political establishment and hydrocracy. Yet, capacity building is one of the weakest mechanisms of foreign development aid. Riddell (2014, p. 9) points to successive evaluations that show aid has generally not been successful in its efforts to build capacities, encourage the retention of high level skills and strengthen public institutions in a sustainable manner. The World Bank in particular has acknowledged that its results in this area have been extremely disappointing (Riddell 2014, p. 9). This is an admission because, as discussed in Chapter 4, capacity building is an increasingly important aid function for the World Bank. Although the World Bank has in the past been seen as a 'serious, engaged partner only when in the lead and in control' (Evans et al. 2015, p. 201) in recent years it has been deliberately shifting toward a facilitating role in bringing together diverse stakeholders to cooperate on a common agenda (Evans et al. 2015, pp. 190–191).

In the Ganges–Brahmaputra problemshed, the 'common agenda' that the World Bank is pursuing is that of transboundary water cooperation. Yet, there is a disconnect between the main objective of SAWI and the 'results' that are expected. The overarching SAWI Program Development Objective is '[t]o increase regional cooperation in the management of the Himalayan River systems to deliver sustainable, fair and inclusive development and climate resilience' (World Bank 2014b. p. 36). Yet, the outcome indicators linked to this goal are distinctly short-term, quantifiable and with uncertain bearing on the sustainability of increased regional cooperation. The outcome indicators for this objective are:

- Number of existing or new bilateral or multilateral governance processes that support cooperative water management that have been informed by SAWI activities
- Value of investments secured though bilateral or multilateral governance processes and that have been informed by SAWI activities
- Quality of the planning processes underpinning new investments in terms of either
    - the breadth and strength of stakeholder consultation,
    - a stronger technical basis for investment designs, and/or
    - the pace at which investment designs are agreed.

(World Bank 2014b, p. 36)

In other words, a long-term, transformational goal is expected to be reached through quantitative improvements in stakeholder process and monetary investments rather than qualitative improvements to those processes or investments. The progress toward the other goals of SAWI is measured by similarly short-term indicators (World Bank 2014b. pp. 36–42). Mayne (1999, p. 7) warns that programme managers need to be realistic about the outcomes of their policies, and the accuracy with which effectiveness can be measured. But the language of the development sector in South Asia is, according to Prasai (2014), unrealistic because it focuses on outputs and outcomes, forgetting that progress is a long-term struggle and can be nebulous and unquantifiable.

There is little acknowledgement within the planning and documentation of these international interventions that the improvement of transboundary water governance in the Ganges–Brahmaputra problemshed is a long-term objective, not easily measured, and one that cannot readily be brought about by a single programme or project. Indeed, progress on cooperation involves, in Prasai's (2014) words, slowly but steadily 'chipping away at small parts of the problems' (for example, the culture of secrecy) by many organisations, small and large, working in concert; change cannot be brought about in one donor funding cycle and neatly quantified by static performance indicators.

The Abu Dhabi Dialogue (ADD) had a 'ten-year vision' for what Vaidya and Sharma (2014, p. 4) describe as a 'cooperative and knowledge-based partnership of states for fairly managing and developing the Himalayan river systems to bring economic prosperity, peace and social harmony, and environmental sustainability from the source to the sea'. Yet, a decade may be too short a timeframe for achieving goals as profound and far-reaching as these. As Earle et al. (2015, p. 41) point out, 'the World Bank has failed in its hope to bring South Asian countries within a common water-sharing framework' in the lifespan of the Abu Dhabi Dialogue process.

The problem of attribution is also not clearly dealt with in the grey literature from international actors in the Ganges–Brahmaputra problemshed. As discussed above, there are often alternative plausible explanations for certain outcomes in addition to the explanation that the aid programme being evaluated was effective in reaching its objective. Mayne (1999, p. 13) suggests that dealing with these alternative explanations explicitly is often the best way of buttressing an argument in favour of the programme's impact. This means identifying the most likely alternative explanations and discussing arguments for and against the evidence for each in order to arrive at the most likely one (Mayne 1999, p. 13). Yet there seems to be very little information in the publicly available documents from SAWI and other international actors about dealing with the problem of attribution, either with Mayne's method or otherwise.

Foreign aid is driven by foreign agendas

The final structural limitation on donors and international organisations in the Ganges–Brahmaputra problemshed is that international development projects are more often than not guided by foreign interests rather than the inherent worth or necessity of the project in question. As Riddell (2014, p. 1) points out, 'the allocation of official aid is still influenced by the short-term political, security and commercial interests of donor governments'. Chapter 4 illustrated how the World Bank has been driven to intervene in the Ganges–Brahmaputra problemshed by considerations other than the limited transboundary water cooperation between India, Nepal, Bhutan and Bangladesh.

Donors regularly switch the countries to which they provide aid, argues Riddell (2014, p. 24); donors also annually vary the amounts of aid they provide. This variability and unpredictability in the system of foreign aid affects effectiveness of development projects and programmes, and exacerbates the problem of short-term vision because there is little certainty and continuity for long-term initiatives. Riddell (2014, p. 24) adds that '[f]or some donors, commercial interests, for example through aid-tying, have also played an important role in determining aid flows. These factors influence which country will receive aid, the amounts provided and the period for which aid is given'.

Which country receives aid and what for is also influenced by what is happening in the wider economy and by donor attitudes. These, writes Riddell (2014, pp. 10–11), have sometimes been assessed as too cautious. For the Ganges–Brahmaputra problemshed that means international actors are keen to fund transboundary water cooperation programmes as long as water security, good governance, sustainable development, non-traditional security threats and so on remain at the forefront of global discourses. Once that is no longer the case, it is possible that interventions for improving water interaction will be discontinued, and they may be discontinued irrespective of whether or not sustainable progress has been attained.

12 Insights for international actors in the Ganges–Brahmaputra problemshed and beyond

In light of the obstacles to positive transboundary water interactions emerging in the Ganges–Brahmaputra problemshed discussed in Chapters 7, 8, 9 and 10, what can be done now to increase cooperation between India, Nepal, Bhutan and Bangladesh? Is there a role for international actors? Those are the central questions of this chapter.

This chapter suggests some alternate approaches to improving transboundary water interaction in the Ganges–Brahmaputra problemshed. These approaches are based on the insights of riparian stakeholders. This chapter begins with a discussion of how addressing historical grievances and the complexities of political reality could create a stronger foundation for the development of transboundary water cooperation. Similarly, countering India's hydro-hegemony and 'levelling the playing field' could dispel some resentment and build trust between riparians. This is discussed in the second section, while the third examines how international organisations and foreign aid donors could increase political will within the Ganges–Brahmaputra problemshed for addressing transboundary water issues. The fourth section examines the possible effects of de-securitising water and taking a water justice approach to transboundary interactions, while the fifth advocates framing issues according to the 'problemshed' rather than the hydrological watershed. The sixth section argues for the strengthening of water sharing institutions, while the final section discusses how to shift away from narratives of water scarcity and supply-side solutions.

To some extent, international actors are already doing these things, either explicitly or implicitly. The purpose of this chapter is not to highlight the shortcomings of the World Bank and other foreign aid donors in South Asia. Rather, it is to emphasise that by tweaking existing strategies to better focus on the suggested approaches, they may somewhat improve their effectiveness. Of course, there are no silver bullets in transboundary water governance, and as was made clear in Chapter 11, it is near impossible to link cause and effect over the long term in international development. Still, room for improvement exists, and this chapter provides some recommendations for international actors.

Addressing the political context and historical grievances

One Indian water security analyst claims that the South Asia Water Initiative (SAWI) has been successful at bringing about a conversation about transboundary waters, but not at bringing about cooperation because little effort has been made to understand what the most pressing problems are (Anonymous 6, 2014). Defining or prioritising water conflicts is the first step to improved transboundary water governance.

The first task for international actors is a scoping one. What is at the crux of the problem with transboundary water governance in the Ganges–Brahmaputra problemshed? Chokkakula (2012, p. 4) contends that 'the search for solutions is obscured by our failure to understand the problem itself completely: the problem lies in lack of acknowledgement of the critical significance of politics in emergence and recurrence of disputes'. This is a problem in the Ganges–Brahmaputra problemshed: the political roots of water disputes are not targeted by international actors here.

Politics at all levels – local, national, regional, global – affects transboundary water interaction by shaping dynamics between riparians and other stakeholders, and by presenting priorities for the state that may not be related to water. As Julien (2012, p. 62) notes, hydropolitics is first and foremost about politics, not water. It is therefore the political dynamics in which water management occurs that must be the focus of analysis. At the same time, these dynamics cannot be reduced to simple inter-state power ratios or economic calculation; they must also deal with culture, history and ideology (Julien 2012, p. 62). Understanding and addressing the political dynamics inside which transboundary water management occurs in South Asia, then, may help international actors better deal with the emergence and recurrence of disputes over water.

A large part of the political context in the Ganges–Brahmaputra problemshed is the history of disputes over transboundary rivers that continue to affect current water interaction. Speaking of Asia as a whole, Chellaney (2014, p. 643) posits that unassuaged historical grievances have constricted diplomatic space for building political accommodation and reconciliation. Asian states, he argues, must find ways to overcome this history of antagonism if they are to cooperate over shared rivers (Chellaney 2014, p. 643). In other words, addressing historical grievances is a prerequisite for transboundary water cooperation. As Perlman-Petersen and Wolf (2015, p. 4) argue, addressing past and present grievances is a requirement of market-driven solutions for enhancing transboundary water cooperation. As such, third parties already engaged in the Ganges–Brahmaputra problemshed ought to consider how some of these past grievances – such as those discussed in Chapters 2 and 3 – can be resolved and serve as the foundation for building positive transboundary water interaction.

Building trust and addressing power asymmetry

One of the outcomes of addressing historical water disputes may turn out to be increased trust between the riparian states of the Ganges–Brahmaputra problemshed. As Earle et al. (2015, pp. 41–42) have noted, the trust deficit is particularly high in this region. This is a real barrier to transboundary water cooperation. Chellaney (2014, p. 643) argues that '[w]ithout improved inter-country relations and better trust, Asia's hydropolitics will remain grating'.

Trust is a prerequisite for positive transboundary water interaction. As Asthana and Shukla (2014, p. 274) argue, 'trust building can lead to institutions of regional water governance with a set of rules and mechanisms to mitigate conflict and disputes'. But how to build trust? Asthana and Shukla (2014, p. 274) suggest that a regional information base where all countries can have access to information may reduce data secrecy and thus be a positive pathway to increased cooperation and peace building.

To a certain extent, international actors are already facilitating knowledge sharing in the Ganges–Brahmaputra problemshed with the objective of building trust and weakening the culture of data secrecy in the region (Zeitoun et al. 2016, p. 14). This was shown in Chapter 6. But these efforts have not built trust or shifted power dynamics between riparian states. As Zeitoun et al. (2016, p. 14) write of SAWI, this foreign-led initiative has not transformed existing transboundary water arrangements. The reports and other publications produced by SAWI are, on the surface, an attempt to provide new perspectives on transboundary water governance in this region. These efforts generally seek to build consensus on the importance of regional cooperation over water – but without challenging the fundamental power relations. SAWI is thus taking a reformist approach that is not likely in the future to transform transboundary water arrangements or to resolve the transboundary water conflict (Zeitoun et al. 2016, p. 14). Without addressing the effects of India's hydro-hegemony, internationally led efforts at knowledge sharing are unlikely to change the *status quo* between riparian states.

Given India's hegemonic position in the region, argue Asthana and Shukla (2014, p. 344), 'multilateralism or third-party interventions in a regional waterscape are not only difficult but also impossible' because perceptions of India's domination still remain in the eyes of smaller neighbours. Many consider India to be the regional bully, point out Earle et al. (2015, pp. 41–42), and India's refusal to manage the region's rivers in a multilateral forum reduces the possibility for any basin-based organisational framework to emerge.

How can international actors better address this situation of India's hydro-hegemony and the mistrust of Nepal, Bhutan and Bangladesh? Zeitoun and Jägerskog (2009, pp. 12–13) suggest that power asymmetry in transboundary water settings can be confronted in two ways: it can be strategically influenced or it can be challenged. There are two general ways to influence power: by deriving positive-sum outcomes, and by creating conditions that 'encourage basin bullies to transform into basin leaders' (Zeitoun & Jägerskog 2009, pp. 12–13). The negotiation of 'win–win' outcomes may satisfy all parties and thus render power asymmetry irrelevant, while bullies may be less likely to force an inequitable arrangement if, for example, they are held accountable to an objective standard, or risk being 'named and shamed'. As Zeitoun and Jägerskog (2009, pp. 12–13) point out, 'not all hegemons are leaders. This makes the role of third-parties all the more crucial.' In the Ganges–Brahmaputra problemshed, third parties such as SAWI ought to encourage India to more consistently take the position of leadership.

There are also two general ways to challenge power: third parties can either level the 'players' or level the 'playing field' (Zeitoun & Jägerskog 2009, pp. 12–13). To help level the players, suggest Zeitoun and Jägerskog (2009, pp. 12–13), capacity-building programmes can improve the technical, administrative or negotiation abilities and empower non-hegemons by increasing their bargaining power. Non-hegemons may thus be able to play a more effective role in transboundary water interactions. This approach also helps them generate their own solutions to collective action problems. As the formerly weaker actors take on more responsibility, they gain respect and power (Zeitoun & Jägerskog 2009, pp. 12–13).

As for levelling the 'playing field', Zeitoun and Jägerskog (2009, pp. 12–13) show that while transboundary water arrangements based on coercive terms determined by a hegemon cannot endure in the long term, the fact remains that the only rule of the game in many basins is the 'law of the jungle'. Effective legislation and a strong regulatory context, they argue, already level the playing field at the sub-national level in many countries. At the international level, the 1997 UN Convention on the Law of Non-Navigational Uses of International Waterways (UNWC) serves as a useful guide to fair water-sharing (Zeitoun & Jägerskog 2009, pp. 12–13). International actors in the Ganges–Brahmaputra problemshed, then, should keep the hydro-hegemon accountable and assist riparians in strengthening their legislative and regulatory frameworks.

Creating political will

Foreign aid donors and international organisations intervening in the hydropolitics of the Ganges–Brahmaputra problemshed can also assist riparian states in putting transboundary water issues higher on the agenda. This may have the effect of increasing political will for addressing water conflicts, which, as discussed, is a necessary condition for cooperation to develop. As Tiwary (2006, pp. 1686–1687) explains, 'international water conflict resolution requires considerable political will on all the sides … It requires sustained motivation to take into confidence different parties (both inside and outside the nation).' And in the words of Shamir (cited in Julien 2012, p. 62), '[i]f there is a political will for peace, water will not be a hindrance. If you want reasons to fight, water will give you ample opportunities.'

In the Ganges–Brahmaputra problemshed, the need for peace and cooperation is growing as environmental security challenges intensify. Earle et al. (2015, pp. 41–42) note that floods and droughts are likely to become more severe and less predictable as a result of climate change. Basin-based cooperation is needed to address these extreme challenges. Cooperation is needed to augment the irrigation capacity in low-season periods and to decrease flood peaks in the monsoon months (Earle et al. 2015, pp. 41–42). The role of the third parties already engaged in this basin is to continue impressing upon India, Nepal, Bhutan and Bangladesh the importance of collective action as an approach for tackling these increasingly urgent challenges. Continuing with Track II dialogue and other informal processes that bring various stakeholders together is a step in the right direction. But it must be sustained over the long term – 20 years or more – to be effective in increasing regional cooperation over shared rivers.

Taking a water justice approach and de-securitising water

Iyer (2007, p. 199) argues that the idea of regional cooperation is gaining currency in South Asia, and is seen as an obvious good – the question is, cooperation at what level, between whom, and for what purposes? To a large extent, that is determined by hydropolitics and power asymmetry between riparians. As Patrick (2014, p. 70) points out, 'who decides what is just and for whom depends on who or what is excluded or included in the water-allocation decision-making process'. Patrick et al. (2014, p. 2) also notes that sometimes just and unjust outcomes coexist: in some cases a just outcome at one level may cause an injustice at another level for the same or a different set of stakeholders.

Patrick (2014, p. 70) also argues that social inclusion of a variety of stakeholders into the dialogue process is a precondition for attaining just outcomes (while the exclusion of any particular stakeholder group may result in unjust ones). Currently, however, water governance processes in the Ganges–Brahmaputra problemshed are not always inclusive. For instance, Chokkakula (2012, p. 12) shows that existing legal structures in India do not allow for the participation of non-state political actors, such as individuals or groups, in the adjudication of inter-state water disputes. This is problematic because, as Iyer (2010, p. 76) argues, any consultative, interactive approach to conflict resolution must also be an inclusive one: it must include all water users, namely the farmers, industrial establishments, municipalities and individuals. He continues that the present system of inter-state water dispute resolution in India is *de facto* an inter-government system. As such it cannot be said to be one that focuses on water justice outcomes or procedural justice (Iyer 2010, p. 76).

Mohan (2010, p. 10) explains that the political boundaries of states 'often subsume issues that are humane, common and social in nature'. In other words, the needs and rights of individuals tend to be overlooked when water issues are framed at the state level. This is the case in transboundary water governance in the Ganges–Brahmaputra problemshed. As Prasai and Surie (2013, p. 1) point out, consideration of perspectives from a broad range of stakeholders is vital to effective transboundary water governance in South Asia. In particular, it is essential to addressing the social and ecological dimensions. Bringing the discourse on water use, sharing and management closer to affected communities and stakeholders can, they argue, help reduce environmental degradation and thereby diminish the potential for conflict (Prasai & Surie 2013, p. 1). But, they continue, 'the lack of regional cooperation and the absence of local and sub-national perspectives continue to prevent sustainable development and management of transboundary water resources for livelihood improvement, food security, poverty reduction, and effective adaptation to climate change' (Prasai & Surie 2013, p. 1).

Prasai and Surie's point about the absence of diverse perspectives echoes Asthana and Shukla's (2014, p. 273) argument about genuine (as opposed to merely rhetorical) inclusivity. Asthana and Shukla (2014, p. 273) advise that good governance should be more than just rhetoric. 'In most cases,' they argue, 'participation of stakeholders begins only when the decision is already announced, after which stakeholders engage in activist roles and resort to collective action rather than being part of the planning process.' Instead, policy should be inclusive of other voices in the formation process, engaging civil society, and moving beyond the top-down exclusionary approach (Asthana and Shukla 2014, p. 273).

Patrick (2014, p. 70) points out that social inclusion is the basis of water justice. If social inclusion is achieved, she argues, then the variety of perspectives on what and why water is needed by the different users for different uses can inform policy and lead to just outcomes. On the other hand, if certain opinions, needs or world-views are excluded, then injustices may develop.

International actors, then, ought to be actively inclusive of diverse perspectives. This does not require any additional funding or fundamental change to the way foreign-led interventions are conducted. Indeed, inclusivity is in line with the tenets of good governance and consistent with Track II dialogue.

Focusing on water justice is also a legitimate approach for addressing water conflicts. As Zeitoun et al. (2014b, p. 188) point out, transboundary water justice is influenced by many of the same dynamics that impact transboundary water conflict and cooperation. These dynamics include power asymmetry, the construction of scale to promote agendas, and the use of discourse to frame tradeoffs and legitimacy (Zeitoun et al. 2014b, p. 188). They also argue that water disputes linger so long as arrangements over distribution and control of transboundary water resources is perceived as unfair by any side, even the weaker one. In that situation, the temptation is to reach for what Zeitoun et al. (2014b, p. 184) term 'practical and expedient ideas'. More novel, difficult, or long-term approaches for addressing the causes of perceived injustice tend, therefore, to be ignored or delegitimised (Zeitoun et al. 2014b, p. 184). Benefit sharing falls into this category of seemingly equitable distribution that may nonetheless not mitigate perceptions of injustice.

Whether water is distributed equitably is often determined by the most powerful actors. Zeitoun et al. (2014b, p. 185) point out that politics and powerful agendas often control the space in which water sharing arrangements evolve. The consequence is that processes that appear procedurally fair may lead to highly asymmetric outcomes. They warn that recognition and participation may be necessary but not sufficient conditions for 'just' transboundary water outcomes, as focusing on process may in effect displace the basic condition for (egalitarian) justice, from equitability of outcome for different actors, to simply their inclusion in an apparently fair process (Zeitoun et al. 2014b, p. 185).

Conca (2006, p. 3) adds that too often hydro-diplomacy and other internationally driven initiatives for improving transboundary water governance are limited to dividing water supplies equitably between nations and reducing the potential for international conflict. As such they do little to address the human security of people living in the basin.

There are no easy solutions to establishing equitable outcomes for all actors at all times, and addressing the human security of the people affected. But Hill (2015, p. 729) proposes that one starting point is the de-securitisation of water issues. He argues that '[t]o desecuritise water in the Himalayas is to open up this dialogue and in so doing create the conditions for just and sustainable development'. He continues that the securitisation of water may detract from discussion about the just and sustainable distribution of water, especially in regard to the poorest and most marginalised parts of society within these countries (Hill 2015, p. 731).

Ultimately, international actors should work toward de-securitising water, while shifting the emphasis away from state-centric understandings of security to human security. Hill (2015, p. 740) suggests that understanding non-traditional security in a way that incorporates a nexus of food, water and energy could make this easier. Regional integration of upstream and downstream communities in the Himalayas must be done, he argues, in ways that achieve integration across the many ecosystems of the Hindu Kush Himalaya (Hill 2015, p. 740). That, in turn, requires reconceptualising the region in terms of a problemshed – not just a hydrological basin.

Seeing the problemshed, not just the basin

Sharma (2014) suggests that taking a basin-wide approach is one way of increasing cooperation. He argues a basin-scale approach by definition requires close coordination with all the countries sharing the basin, such as Nepal and Bangladesh, so that the interests of both upstream and downstream users are taken into consideration (Sharma 2014). Although a basin-wide approach might not always be appropriate for all water management issues, and the inclusion of all riparian states could slow down decision-making processes, the point remains about developing a shared understanding of managing water resources.

Thakkar (2014) points out that SAWI has been relatively successful in taking a basin-wide approach. He sees the inclusion of China in the foreign-led multilateral forums – the ADD and SAWI – as particularly commendable. This is important because as Earle et al. (2015, pp. 41–42) have noted, the advent of China in the Ganges–Brahmaputra–Meghna basin's hydropolitics has brought new complications and uncertainties, much like it did in the Mekong and Nile basins. In this sense taking a basin-wide approach is useful for inclusivity, trust-building and creating a shared sense of mission in regard to the Ganges and Brahmaputra Rivers.

At the same time, taking a basin-wide approach is not always appropriate to all transboundary water governance issues. Swain (2011, p. 696) shows that a basin-wide approach was not successful in the Nile Basin Initiative, on which the Abu Dhabi Dialogue was modelled. He adds that a sub-basin approach may have had more of a chance of successfully increasing transboundary water cooperation in the Nile basin (Swain 2002, pp. 293–308). The basin-wide approach requires consensus from all riparian states, and that can be an impediment for taking action on certain issues, especially ones that do not affect the basin as a whole. Moreover, as Islam and Susskind (2013, pp. 51–52) point out, hydropolitics frequently influence – and are influenced by – factors outside the boundaries of watersheds. This is also the case in the Ganges–Brahmaputra–Meghna basin.

Considering the Ganges–Brahmaputra problemshed as a separate and distinct unit of water management may be a more functional scoping of transboundary water governance issues than considering the Ganges–Brahmaputra–Meghna basin as a whole. To recap, a problemshed is 'a geographic area that is large enough to encompass management issues, but small enough to make implementation feasible' (Islam & Susskind 2013, p. 321). Islam and Susskind (2013, p. 42) argue that at the heart of many of our water management difficulties lies the fragmented or bounded view of water as a 'natural object', or at other times, as a 'societal issue' or at still other moments as a 'political construct'. The components of a water resource management puzzle can fit together in so many different ways that it is practically impossible to effectively use traditional systems engineering approaches to resolve complex water governance problems.

These complexities may be better accounted for in the concept of a problemshed or policy-shed, which takes into account combinations of natural, societal and political considerations (Islam & Susskind 2013, pp. 51–52). Instead of thinking about managing systems that are bounded and made up of components that interact in predictable ways, Islam and Susskind (2013, p. 42) suggest that international actors may be more effective in increasing transboundary water cooperation if they begin to think in terms of complex water networks.

International actors may also find it helpful to think in terms of problemsheds rather than basins when working to increasing transboundary water cooperation in South Asia (while ensuring that the hydro-hegemon does not perpetuate the *status quo* of inequitable bilateral agreements). This may, in the words of Prasai and Surie (2015, p. 32), promote 'alternative imaginations in transboundary water cooperation'. They argue that reconceptualising the value of water and the scope of water negotiations can reshape what transboundary water cooperation looks like. Prasai and Surie (2015, p. 32) provide two examples of what they mean. One is the 'devolution of negotiating authority to the local level where stakes are real and the prospects of making reasonable compromises on cooperation frameworks are greater'. The second is seeing 'beyond immediate calculations of allocated flows to include other ecological, geopolitical and economic benefits' (Prasai & Surie 2015, p. 32). These non-water related benefits may occur beyond the basin, and thus are more likely to be identified through a problemshed approach.

Understanding that water is a socially constructed concept may help reveal previously overlooked benefits, and to de-securitise water. As Hill (2015, p. 738) points out, securitisation of water is a social construction that reflects prevailing power relations within and between states rather than an objective existential threat. As such, a nuanced understanding of socio-political dynamics and their impact at a number of levels clearly has a great deal to contribute to our understanding of the emerging situation in the Ganges–Brahmaputra–Meghna basin. That nuanced understanding is currently missing in the region. instead, notes Hill (2015, p. 738), water security is framed in terms of a contest over the use of a river, discussed in terms of a quantifiable volume of flow. Indeed, the debates between and within countries of the region are frequently defined in this way, such as the long-running issues over the Farakka Barrage, or the debate around the Teesta River.

There are many epistemological and ontological assumptions underpinning this alternative approach. The river is conceived as a resource detached from and managed outside its broader environments and cultural context. But this conceptualisation, points out Hill (2015, p. 738), neglects the wider, and equally important, riparian ecosystems. This, in turn, limits how the benefits to and from the river are perceived. Hill urges international organisations to think beyond the Himalayas in terms of its capacity to be harnessed for energy and agriculture. Instead, he suggests that those involved in transboundary water governance should explore the function of rivers as important in the livelihoods of a great number of people and as a provider of ecosystem services. Such a conceptualisation allows for a very different picture of the costs and benefits of harnessing transboundary water resources (Hill 2015, p. 738). Taking a problemshed view, then, rather than a watershed approach, may help international actors illuminate these 'hidden' costs and benefits of transboundary water.

Strengthening water sharing institutions

In 2014, the Berlin-based consulting firm Adelphi released a report written by Benjamin Pohl et al. called *The Rise of Hydro Diplomacy*. Pohl et al. (2014, p. 35) claim that the most important role for third party interventions is in strengthening water governance institutions. They argue that to decrease conflict and increase cooperation, foreign policymakers must:

- exert political leadership in fostering intra-basin cooperation and integration;
- connect and reinforce appropriate institutional structures for coordinated and cross-sectoral engagement;
- strengthen the diplomatic track of transboundary cooperation;
- invest more in training and capacity-building;
- expand efforts to build confidence in shared basins; and
- improve water-related crisis response and conflict resolution mechanisms.

(Pohl et al. 2014, p. i)

Regional water governance institutions may increase transboundary water cooperation by creating a shared understanding of rivers. The water resources of the Hindu Kush Himalaya are a shared resource in many ways, point out Vaidya and Sharma from ICIMOD. Transboundary rivers are shared between upstream and downstream users, not just between countries, but also between people and nature. This shared nature of water brings the promise of multiple benefits, but also the threat of increased conflict. To realise the benefits of water resources, argue Vaidya and Sharma (2014, p. iv), it is essential that there is a shared understanding of the resource.

Asthana and Shukla (2014, p. 274) have several suggestions for how to increase regional awareness that rivers can be better harnessed through collective efforts. First, building trust between riparians is key. This has already been discussed. Second is establishing a set of rules and mechanisms to mitigate conflict and disputes. Third, they propose a regional information base where all countries can have access to information and take adequate steps to mitigate damage, to remove problems of data secrecy (Asthana & Shukla 2014, p. 274). Asthana and Shukla (2014, p. 344) would like to see the establishment of a permanent, politically neutral forum similar to SAWI but not run by international actors. But any such forum must overcome historical and regional biases and perceptions of India's hydro-hegemony. It must be an institution that can fund regional experts and epistemic communities in a neutral space and unbiased environment (Asthana & Shukla 2014, p. 344).

The effect of India's hydro-hegemony on the effectiveness of water sharing institutions in the Ganges–Brahmaputra problemshed has been a point of contention. Hill (2013, pp. 255–256) argues that while some commentators feel that these existing institutions (such as the Mahakali Commission between India and Nepal, or the Joint Rivers Commission between India and Bangladesh) are in the main effective, much of the commentary dismisses the importance of these bodies. Instead, the relationships between the basin countries are 'dictated by India's overwhelming economic and political dominance' (Hill 2013, pp. 255–256).

Strong river basin organisations could to some extent counter India's hydro-hegemony. Transboundary water sharing institutions can, according to Earle et al., help to reduce uncertainty surrounding Ganges–Brahmaputra riparian relations. Moreover, such institutions may shift the focus from disconnected and short-term interactions of the riparian states into a continuous relationship that has scope for future routine gains, claim Earle et al. (2015, p. 41).

Feldman (2007, p. 3) suggests that water sharing institutions should be based on an adaptive management approach, which requires four things: 1) an organisation design that permits policymakers to recognise mistakes; 2) an ability to monitor and measure change, especially environmental change brought about by prior decisions; 3) a capacity to adopt mid-course corrections, and 4) an ability to apply what's been learned to more complex challenges. The World Bank and a number of donor countries tried to foster momentum for just such a regional organisation (Hill 2013, pp. 255–256). But as Earle et al. (2015, p. 41) point out, to date all these expectations have not ventured beyond political statements and setting up lofty goals.

But the World Bank is not the only international organisation working to establish or strengthen water sharing institutions in the region. Crow and Singh (cited in Colopy 2012, p. 319), for example, have suggested that the International Centre for Integrated Mountain Development (ICIMOD) might even play a larger role in the region. It could serve in some capacity as a water regulatory institution, monitoring the health of river basins and overseeing participating nations' adherence to international agreements. They have also argued for the establishment of a regional organisation that they call the Himalayan Authority for Water Services and Environmental Co-operation. To these scholars, it is only through such a body that genuine regional co-operation might be possible (Hill 2013, pp. 255–256).

Establishing, strengthening and maintaining river basin organisations is not easy. Hill (2015, p. 732) argues that while there is much potential for new political spaces to be opened up in the Ganges–Brahmaputra problemshed, this will be a long and unenviably difficult process given how politically contentious many parts of the region already are. There are no guarantees of any success in this regard. He also shows that many of these initiatives have, in the past, been faltering. Ultimately, they have had little success in changing the way riparian states think about transboundary water resources (Hill 2015, p. 732).

Given the difficulties of establishing river basin organisations, another path to greater transboundary water cooperation could be forged by civil society. Hill (2013, pp. 255–256) argues that a multilateral institution for transboundary water governance must go beyond Track I (government to government) diplomacy to create space for networks of civil society and people's movements from across the region. Currently, however, civil society in South Asia is limited (this was discussed earlier in Chapter 8). As Hill (2013, pp. 255–256) shows, NGOs in the region tend to be preoccupied with donor-related developmentalism, rather than advocacy. And they are constrained in the extent to which they can express dissent. Another challenge for civil society is the politically fragmented nature of the subcontinent. It makes it very difficult for pan-South Asian networks to regularly meet and forthrightly take up issues associated with trans-boundary water resources (Hill 2013, pp. 255–256).

Despite these challenges, alliances are beginning to form between civil society groups across the region, although as Hill (2013, pp. 255–256) argues, their influence on broader policy debates is still hard to discern because they are relatively new. But, he adds, if such bottom-up initiatives that urge for greater transparency in hydro development in the region were to be incorporated into regional institutions, then it is conceivable that water resource development might proceed in a way that is more equitable, transparent and sustainable than that which currently prevails (Hill 2013, pp. 255–256).

Moving beyond narratives of scarcity and supply-side solutions

Lastly, international actors may be able to use their water governance programmes for more than just supply-side solutions. As Asthana and Shukla (2014, p. 271) point out, there needs to be a shift in emphasis on demand management practices in South Asia and further incentives for conservation of water resources. Chellaney (2011, p. 297), whose focus is on water conflicts and security, agrees that a new strategic approach is needed. The approach he calls for centres on water conservation and recycling, efficiency and productivity gains, and, more broadly, integrated resource management that involves all basin states. Only collaborative paths that embrace all stakeholders and break free from the business-as-usual approach, he argues, can help unlock equitable outcomes (Chellaney 2011, p. 297).

Value creation is necessary to move from competition to non-zero-sum thinking but it is difficult when the negotiations arise out of fear of scarcity. This is because conceptualising water as a scarce resource precludes it from being a flexible resource. Islam and Susskind (2013, p. 131) point out that too many water conflicts arise out of parties fearing they will not have enough water. That fear, in turn, skews thinking on water governance problems toward supply-side solutions.

The supply-side approach to water resource management focuses on availability – 'creating' more water. This, however, can lead to a destructive, self-perpetuating cycle. As Iyer (2010, pp. 79–80) explains, there is always a demand for more water, and this demand is more often than not met by big dams, canals and long-distant water transfers. These in turn generate new conflicts (Iyer 2010, pp. 79–80). Hill (2015, p. 741) agrees and adds that hydropower development across the region is constrained by the overwhelming dominance of the supply-side hydrological paradigm.

One way in which this dominant water governance paradigm could be flipped is by reframing a problem of 'insufficient water supply' as one of 'excessive water consumption'. Patrick et al. (2014, p. 11) argue this may lead to solutions that favour the way in which water is allocated, used and consumed, rather than technological solutions that 'create' more water.

In addition to the focus on supply-side solutions to water scarcity problems, Prasai and Surie (2013, p. 3) argue that water governance policies in South Asia tend to be state-centric, technocratic, exclusionary and nationalistic. Often these policies do not adequately recognise the human, ecological and social costs of water allocation and their implications. Furthermore, they claim, governments in South Asia have traditionally monopolised the right to articulate public interests on water resources and have continued to show a deep resistance to parting with that tradition. This situation has been exacerbated by collusion among politicians and domestic and foreign investors involved in water-related infrastructure projects (Prasai & Surie 2013, p. 3). This results in more engineering and supply-side solutions to water governance problems rather than innovative policy and equitable distribution.

For the foreign investors, then, and other international actors in the Ganges–Brahmaputra problemshed, there exists an opportunity to influence the governance of the region's rivers to better recognise the human, ecological and social costs involved, especially in infrastructure and hydro-engineering projects. They should lead by example.

To some extent, the World Bank is beginning to do this. The Ganges Strategic Basin Assessment, for example, is clear that strategies other than hydro-engineering will have to complement or maximise the benefits of water storage in the region. The study also puts forth the argument that there are many attractive alternatives to meeting the region's water needs; constructing new water storage infrastructure is not the only option. This is a novel and important argument in a water governance context dominated, as discussed in Chapter 2, by the prevailing preference in the region for engineering solutions to water supply policy problems.

Engineering solutions tend to be justified on financial and economic grounds. But, as Hill (2015, p. 741) shows, in the Ganges–Brahmaputra–Meghna basin, hydropower development functions to channel the most economic benefits to a comparatively small group of people. At the same time, hydropower development ignores a range of economic, social and ecological issues that impact upon the livelihoods of many people who depend upon these rivers (Hill 2015, p. 741).

Like the supply-side paradigm, he argues that 'the neo-liberal approach overlooks the significant social, economic and political issues associated with infrastructure-led growth' (Hill 2015, p. 729). Not taking into account these social, economic and political factors can lead to inequitable outcomes and perceptions of injustice. As has been discussed above, perceptions of injustice are the root of water disputes. As such, moving beyond narratives of supply-side, engineering solutions and neo-liberal justifications of infrastructure-led development may promote positive transboundary water interactions.

Conclusions

In 2011 Swain summarised the effect of donor-led, basin-wide cooperative frameworks in the Nile Basin. Six years later, his conclusions can well be applied to foreign-led initiatives into transboundary water cooperation in the Ganges–Brahmaputra problemshed: 'The so-called basin-based cooperative framework in the Nile was initiated with the active involvement of international donor agencies, and it still survives with their help and assistance.' But after more than a decade of existence, the Nile Basin Initiative receives very little contribution from the riparian states and this, argues Swain (2011, p. 698), is 'exposing the lack of interest of the basin states towards joint management of the shared river resources'.

The Abu Dhabi Dialogue was modelled on the Nile Basin Initiative (Earle et al. 2015, p. 41), and as in the Nile basin, the mission of foreign-led interventions to increase transboundary water cooperation in the Ganges–Brahmaputra problemshed has been limited and weakened by several factors. Some of these factors lie in the sheer complexity of the transboundary water governance issues – as well as the enormity of other, non-water related political priorities – in the region. There are competing issues for political will, and transboundary water cooperation is relatively low on the agenda. This was discussed in Chapters 2 and 3.

Other factors limiting the potential for cooperation between India, Nepal, Bhutan and Bangladesh lie in a long history of grievances and mistrust between them. At the same time, foreign-led interventions are not directly targeting the lack of trust between riparians but rather are driven by their own agendas and global narratives; good governance, non-traditional security threats, sustainable development, water security, climate change resilience and so on. These were discussed in Chapters 4 and 5. Chapter 6 outlined the key goals of the World Bank and other international actors in this space. There are, however, many challenges for them. There is little buy-in from local stakeholders for foreign-led interventions, zero-sum attitudes to water sharing prevail in the region, and India's hydro-hegemony prevents the development of multilateral solutions. These were discussed in Chapters 7, 8 and 9. Lastly, the approaches favoured by international actors may be inadequate to bring about a significant improvement in transboundary water interaction. Track II dialogue is a necessary condition of cooperation, but it is not a short-term solution. Yet, the relatively short donor funding cycles and other weaknesses in the system of foreign aid, may be mismatched with the long-term, complex and unpredictable process of improving transboundary water governance. These factors were discussed in Chapters 10 and 11.

What, then, can we learn from the World Bank and other international actors working to increase transboundary water cooperation in the Ganges–Brahmaputra problemshed? That is the central question of this, the final chapter. As Swain's remarks above illustrate, the urgency 'to meet the new challenges that might arise from climate change' is increasing, but workable solutions are not necessarily forthcoming from either the international donor community or the riparian states themselves (Swain 2011, p. 698).

This last chapter recaps the story of international actors in the Ganges–Brahmaputra problemshed. Why are they engaged here? Why do the water interactions in this region need to improve? First, there is a summary of the transboundary water governance between India, Nepal, Bhutan and Bangladesh. Then there is a discussion of the challenges international actors face in trying to increase transboundary water cooperation and how they may be able to overcome them. Although limited, there are some positive effects that third parties may achieve as part of what one riparian stakeholder described as a slow but steady 'chipping away' at the problem of transboundary water interactions.

There are many pressing transboundary water issues but little cooperation between India, Nepal, Bhutan and Bangladesh

There are numerous issues for transboundary water governance in the Ganges–Brahmaputra problemshed. For a start, it is a region particularly vulnerable to hydrological and ecological shocks; it is one of the most densely populated, rapidly growing and poorest parts of the world. The increasing demand for food, energy and water is putting tremendous stress on the problemshed's already strained natural resources. At the same time, governance of these resources is generally weak and marred by suspicion, mistrust and long-standing grievances between riparians. This section outlines some of the contextual factors of transboundary water governance in the Ganges–Brahmaputra problemshed, and summarises what international actors are doing here and why.

The Ganges–Brahmaputra problemshed is vulnerable to water related shocks

The Hindu Kush Himalayan region, of which the Ganges–Brahmaputra problemshed is a part, has one of the largest bodies of ice outside the polar caps, and the mountains are often referred to as the 'water towers' of Asia. These mountains are the major source of both surface and groundwater in the region during the dry season, and have the potential to play a vital role in energy security; if properly harnessed, hydropower could transform the lives of the people living in these river basins. But, as Vaidya and Sharma (2014, p. 3) point out, in the wet monsoon season, the contributions of meltwater and rainfall coincide, creating a situation of too much water in the wet season and too little water in the dry season. This is exacerbated by the lack of water storage facilities and is already posing a serious threat to livelihoods and economic development in the region.

The Ganges–Brahmaputra problemshed is an area of both plenty and scarcity. Both extremes are likely to be exacerbated by the effects of climate change. As Hill (2015, p. 739) notes, the phrase 'too much and too little water' is an apt description of the challenges in the Hindu Kush Himalaya, with droughts and floods becoming more commonplace.

The Ganges–Brahmaputra–Meghna basin is one of the richest basins in the world in terms of natural resources (hydropower, fisheries, forestry, irrigated agriculture, navigation, environmental amenities, tourism, minerals etc.). Yet India, Nepal and Bangladesh are some of the poorest in the world and sometimes called 'the poverty triangle' (Dinar et al. 2007, p. 251). It is also a densely populated region; approximately 630 million people in an area of 1.7 million square kilometres (Vaidya & Sharma 2014, p. 17).

India today has to support over 17 per cent of the world's population on just over 2 per cent of the world's land area, while Bangladesh supports about 2.27 per cent of the world population on 0.11 per cent of the world's land area. These ratios are bound to widen by mid-century (Wirsing et al. 2013, p. 14). This population growth, combined with rapid urbanisation and fast-paced economic development means that demand for water already outstrips supply (Kumar & Furlong 2012, pp. 12–13). Rapid urbanisation alone 'is driving increased water demand both for municipal supply and for the industrial and agricultural products in demand in cities' (Chellaney 2014, p. 622).

Wirsing et al. (2013, p. 14) identify this 'water scarcity and demographic plenty' as one of the circumstances that make the Ganges–Brahmaputra problemshed particularly vulnerable. The others include flooding/drought, damming of the rivers, faltering hydro-diplomacy and political turmoil. Climate change is also on their list (Wirsing et al. 2013, p. 14). The compounded effects of rising temperature, flooding, sea level rise, loss of wetlands and cyclones (Evans 2015b, p. 80) will leave densely populated 'mega-deltas', such as the Ganges–Brahmaputra–Meghna basin particularly at risk (Evans 2010, p. 4). Other potential effects of climate change include complications with spatial and temporal rainfall variability, evaporation rates, and temperatures in different agro-climatic zones and river basins (Asthana & Shukla 2014, p. 139). These, in turn, will likely affect 'agricultural production and food security, ecology, biodiversity, river flows, floods and droughts, water security, and human and animal health' (Vaidya & Sharma 2014, p. 17).

The frequency and intensity of floods in the river basins is expected to increase as a result of an increase in precipitation during the monsoon season (Vaidya & Sharma 2014, p. 3). Changes in climatic conditions also affect diseases transmitted with water, via such vectors as mosquitos (Werz & Conley 2012, p. 20). Moreover, the countries in the Ganges–Brahmaputra problemshed are less resilient to water-related disasters than other Asia-Pacific countries, which creates a further challenge for transboundary water governance and climate change adaptation (Vaidya & Sharma 2014, p. 12).

There are competing transboundary water issues

In addition to these sources of volatility, there are competing issues on the transboundary governance agenda. Each presents unique risks and opportunities, but they are largely interrelated and, as with any wicked problem, addressing one issue in the problem set may adversely affect another. The issues identified by the World Bank as being most pressing are: upstream water storage, water for irrigation, hydropower generation, flood mitigation and sedimentation. (Pollution and groundwater exploitation are also major issues for the riparians, but of lesser consequence for transboundary interaction.)

Water storage is a particularly important issue for transboundary governance because it can be used to even out water supply between where it is comparatively plentiful – Nepal and Bhutan – and where it is scarce – India. In addition to flow augmentation for downstream irrigation, Nepal's storage potential of 88 bcm could bring other mutual benefits, such as hydropower generation, flood control during the monsoon, and flow augmentation for navigation (Onta 2001, p. 106).

It is said that water and energy pose the biggest constraint on India's growth because demand for both is increasing at a rate faster than current capacity can provide (Kumar & Furlong 2012, p. 8). Nepal's enormous hydropower potential could provide a convenient and significant supply of 'clean' energy for India's growing needs, particularly in the north of the country (Onta 2001, p. 11). Yet hydropower production remains largely undeveloped for domestic uses and export (Gyawali 2001, p. 3784).

Although the issue of hydropower generation in the Ganges–Brahmaputra problemshed is the one most often – and most vigorously – debated within the region and internationally, it is water for irrigation that has broader effects in terms of human security: livelihoods and food production. As Dinar et al. (2007, p. 254) point out, agriculture uses half of all freshwater resources in the Ganges–Brahmaputra–Meghna basin, making water supply one of the most significant barriers to economic development. Agriculture, in turn, is vulnerable to drought and flooding.

Floods are a natural part of the hydrology of the Ganges–Brahmaputra problemshed and serve essential ecosystem functions such as groundwater recharge, soil fertilisation and fish stock renewal (Thomas 2012). Flooding also brings nutrient-rich sediment, and rice harvests were segmented in pre-colonial times throughout the year to hedge against flood pattern fluctuations (D'Souza 2007, pp. 263–264). But the understanding of floods as bringing benefits as well as dangers has been lost with the development of the 'command and control' approach to water governance in the past century. The problem of flooding, especially for Bangladesh, is not just one of a surfeit of water, but also a surfeit of sedimentation. Compounding the diversions of Ganges River flow to Bangladesh is the rapid rise in silt deposition rates since the Farakka Barrage was erected upstream in India in 1975. Such sedimentation substantially reduces the river bed's storage capacity and further increases flood risk (Thomas 2012).

These issues of hydropower generation, irrigation, flooding and sedimentation are brought together in the prevailing narrative of freshwater resource development in the region. The prevailing thinking is that there is almost infinite hydropower potential in the Himalayas (especially Nepal) and that it should be exploited for the benefit of all riparians. Planned upstream dams are framed as 'multi-purpose' and promise to bring multiple benefits: economic benefit for Nepal, energy security for northern India, water storage for dry season flow augmentation and irrigation, and flood control throughout the basin. This narrative is based on the dominant paradigm of water management: command and control of nature, and the pursuit of technological and engineering solutions to water related problems.

The World Bank has attempted to challenge some of the assumptions in this narrative through the 2014 Ganges Strategic Basin Assessment. This study is the first to comprehensively model the hydrological, economic and social aspects of this basin, and to provide the foundations for evidence-led policy. The Ganges Strategic Basin Assessment intended to challenge existing misconceptions about how to manage the region's freshwater resources, and commence a new era of cooperative governance based on benefit sharing. The World Bank has taken a scientific approach to mapping out how to optimise the various benefits that can be derived from the Ganges basin. But the response from riparian states was less than accepting. India, Nepal and Bangladesh rejected the World Bank study. The Ganges Strategic Basin Assessment had touched a raw nerve, that of underlying and unresolved grievances between these riparians.

Water interactions between India, Nepal, Bhutan and Bangladesh are not positive

As Dinar et al. (2007, p. 248) note, 'the Ganges has tremendous joint development potential that has not yet been realised by its riparian states. Rather, the Ganges Basin is more popularly known for its rich history of disputes.' Factors in this 'rich history of disputes' include the lack of strong regional identity, the securitisation of water issues, the presence of a hydro-hegemon that struggles to be a regional leader, the reliance on bilateral rather than multilateral approaches to transboundary water governance, and a number of unresolved disputes between India–Nepal and India–Bangladesh. Wirsing et al. (2013, p. 14) also add faltering river diplomacy, and simmering separatist tensions and political turmoil to these conflict-inducing factors.

Uprety (2014) summarises the transboundary water interactions in South Asia by saying that while some riparians claim to be in favour of cooperative, multinational approaches to water governance, collective multilateral attention to the problems of transboundary waters has been rare. The region lacks a collective strategy and bilateralism remains the standard approach to transboundary water governance. Riparian states engage in multilateral water-related activities only half-heartedly, preferring instead to further their own specific interests. This is because, Uprety (2014) argues, of their prevailing mutual distrust, their inability to delink hydropower from water resources, and their reservation against generalising policy to avoid establishing precedent.

South Asia is one of the most poorly governed regions in the world, with inefficient and inequitable deployment of resources, crippling debt burden, social divisions along ethnic and sectarian lines, as well as rampant corruption and vulnerability of civil society organisations (Baqai 2011, p. 27). As Baqai (2011, p. 31) argues, the South Asian state is both strong and weak. On one hand, there is a high degree of 'state-ness', but on the other hand, social disorder and political upheaval is the norm.

The weak solidarity and regional identity in South Asia also means that the region lacks a security community (Jones 2008, pp. 183–193). As Asthana and Shukla (2014, pp. 275–277) show, there is also no sense of collective action for dealing with non-traditional security threats, such as water conflicts. Indeed, they argue that '[n]on-traditional threats are also securitised across borders due to a bitter past, mistrust and hatred that dominates the socio-political structure in these countries' (Asthana & Shukla 2014, pp. 275–277). The securitisation of water continues to be a driver of negative water interaction in the Ganges–Brahmaputra problemshed.

India's hydro-hegemony is a further point of tension for Nepal, Bhutan and Bangladesh (Ray 2008, p. 88). India's relative political, military and economic strength allows it to mitigate the fact of geography that it does not control the headwaters of all the rivers that pass through its territory. India is thus able to influence its upstream neighbours to gain access to their water resources, and its downstream neighbours to overlook transboundary water arrangements that may adversely affect them. This influence, as that of any hydro-hegemon, can at times be constructive and at times coercive (Bhasin 2008, p. 7).

Hydro-hegemony is not inherently a destabilising factor; indeed, a hydro-hegemon could be a strong regional leader that promotes cooperation among riparians. But, as Bhasin (pp. 19–21) argues, India has failed to provide political advice and economic guidance to neighbours without appearing domineering. The perception of India as a sometimes bullying 'big brother' continues to foment resentment, distrust and political tension within Nepal, Bhutan and Bangladesh vis-à-vis the hydro-hegemon.

As Hill (2008, p. 75) shows, the geopolitics of the region is defined by the dominance of India. The question of whether transboundary disputes should be handled bilaterally or internationalised continues to provoke tensions. These tensions, he argues, are likely to intensify as demand for water becomes more acute in the future. Even though all the rivers flowing through India are international and pass more than one country, all the treaties on these rivers are bilateral (Singh 2008, p. 36). India's persistence in establishing strictly bilateral arrangements and not involving the international community in matters of transboundary water governance in South Asia has, argues Hill (2008, p. 60), shaped the transboundary water interaction in the Ganges–Brahmaputra problemshed.

Negative water interactions between the hydro-hegemon and Nepal and Bangladesh are further exacerbated by several unresolved transboundary water issues. Condon et al. (2009, p. 13) argue that Nepal's limited diversion or storage capacity is at the root of water disputes with India. While for Nepal the issue of water storage is largely one of untapped potential, for Bangladesh it is a question of mitigating the negative effects of India's upstream projects. Bangladesh has a seeming abundance of water resources, but most of its annual flow comes in the form of monsoonal floods that quickly flush out to the Bay of Bengal; Bangladesh is a predominantly flat, deltaic country with limited water storage potential (Wirsing et al. 2013, p. 15). At the same time, Bangladesh has not benefitted from water storage and diversion in upstream India.

In contrast, the water interaction between India and Bhutan is largely positive. The cooperative relationship between Bhutan and India over transboundary water resources can be attributed in large part to the kingdom's far-sightedness and political savvy in fostering the non-zero-sum-thinking that allows the interests of both Bhutan and India to be addressed through hydropower development. Bhutan's stance toward the hydro-hegemon can be said to be an instance of 'bandwaggoning', whereby weaker states in a regional system seek accommodation with the local hegemon in order to receive economic and military benefits (Dash 2008, p. 117). Bhutan has been able to negotiate favourable outcomes with India regarding shared waters because, unlike Nepal or Bangladesh, it is a small, politically stable, ethnically and religiously homogenous country that does not have to grapple with many significant domestic priorities (e.g., insurgency and frequent changes of government, as is the case in Nepal) that could dilute political will for transboundary water governance (Hanasz 2015a).

International actors working to increase transboundary water cooperation

The response of international organisations and donors to this situation – environmental vulnerability, numerous issues for transboundary water governance, and less than positive interaction between the hydro-hegemon and the other riparians – has been to frame it as a collective action problem. Colopy (2012, p. 319) points out that intelligent choices about river governance in South Asia seem impossible without regional cooperation. Hill (2013, p. 257) too agrees that 'co-operation between all of South Asia's states for the benefit of the region's citizens … is essential if the multiple crises of water are to be overcome to any significant degree', but worries that transboundary water cooperation in the foreseeable future 'is less likely than continual division and enmity'.

Grey, Sadoff and Connors (all of whom have worked for the World Bank in South Asia) note that there is 'very limited existing transboundary cooperation' in the Ganges–Brahmaputra problemshed, and that although '[f]uture risks are undoubtedly high' they could nonetheless potentially be mitigated through cooperation, especially in the form of joint institutions for information sharing (Grey et al. 2009, p. 17). Cooperative infrastructure development, they argue, could help regulate river flows, to mitigate floods and droughts, generate power and irrigate fields. Meanwhile, cooperative environmental management could enhance water quality and ensure environmental flows for ecosystem health (Grey et al. 2009, p. 17). The underlying assumption in these schemes for increasing transboundary water cooperation is that they require third party support or facilitation. After all, if riparian states had the capacity and political will to mitigate future water security risks on their own, they would have.

In contrast to Grey et al., however, Hill (2013, p. 257) places less emphasis on the role of international organisations in facilitating transboundary cooperation and more on multi-scalar civil society networks. But, he notes, 'at present, such networks are weak, underdeveloped and unable to exert significant influence on state policy and inter-state relations' (Hill 2013, p. 257). This alleged weakness creates an opportunity for international actors to intervene through Track II or Track III dialogue (that is, dialogue between civil society actors).

The framing of water security issues as a collective action problem (one that riparians, especially in the developing world, are presumably ill-equipped to tackle themselves) is not a phenomenon unique to the Ganges–Brahmaputra problemshed but rather it is part of a recent global trend. Since around the year 2000, the new 'sanctioned discourse' of water management has, according to Mollinga and Gondhalekar (2014, p. 180), centred on the ideas of the market, good governance and sustainability. This discourse has been supported by a set of global organisations (notably the United Nations), and actively propagated through international development funding agencies. Thus the framing of the situation in the Ganges–Brahmaputra problemshed as a collective action problem is part of the emerging 'global politics of water', or the globalisation of regional water security issues through international organisations and aid donors (Mollinga 2008b, p. 13).

That is not to say that there is no desire for cooperation and collective action from within the problemshed. On the contrary; scholars from India, Nepal and Bangladesh have been arguing that only collaborative approaches can resolve the water-related disputes in the region. Iyer (2007, p. 199), one of the most prominent and well-respected authorities on water issues in South Asia, points out that the idea of regional cooperation is gaining currency in South Asia – but the question remains, cooperation at what level, between whom, and for what purposes?

Sharma (2014) argues that 'a basin-scale approach would help manage the water resources better' but that such an approach would 'require close coordination with all the countries sharing the Ganga, such as Nepal and Bangladesh, so that the interests of both upstream and downstream users are taken into consideration'. Sharma (2014) also lists various government and non-government agencies, the private sector, civil society groups and the public at large as needing 'to be actively engaged in these efforts' for improving water governance in the Ganges basin. International actors are absent from this list of stakeholders.

Asthana and Shukla (2014, p. 274) also agree that South Asia needs to move forward with a cooperative and participatory approach on river basins and water sharing, and that cooperation is essential to alleviate threats to water security. Their suggested approach for fostering cooperation is that of inclusion and dialogue: 'Policy making needs to be more inclusive of other voices in the formation process, engaging civil society, and moving beyond the top-down exclusionary approach' (Asthana & Shukla 2014, p. 273). For Asthana and Shukla, then, Track II dialogue may be an appropriate approach to improving transboundary water interaction, but it does not necessarily have to be facilitated by international actors.

Even Chellaney (2011, p. 297), a proponent of the idea that Asia may be on the brink of water-related war, agrees that only an approach based on 'integrated resource management that involves all basin states' can prevent water conflicts from worsening. Again, however, international actors are not explicitly made part of the solution to negative transboundary water interaction in the Ganges–Brahmaputra problemshed.

The impetus for foreign aid donors and international organisations in increasing transboundary water cooperation has come not from within the Ganges–Brahmaputra problemshed, but from these external actors. The new 'sanctioned discourse' of good governance and sustainable development, combined with the emergence of the global politics of water, is driving international organisations and donors to invest more and more in the water governance of regions of strategic importance to them. The global focus on climate change has also added, according to Mollinga and Gondhalekar (2014, p. 180), a layer of urgency to 'fixing' transboundary water problems and water security.

South Asia is increasingly portrayed as 'the linchpin of humanity's collective effort' (Matthew 2014, p. x) for collective action problems. Asthana and Shukla (2014, p. 241), for example, argue that '[b]eyond doubt, hydropolitics between India and its neighbouring riparian states is one of the most urgent, complex, and contentious issues facing the regional community in the twenty-first century'. International actors are rising to this challenge. In the Ganges–Brahmaputra problemshed, Colopy (2012, p. 319) shows that '[o]ne sign of this is an effort led by the World Bank called the South Asia Water Initiative, funded by Great Britain, Australia, and Norway'.

SAWI was established in 2009 following the World Bank-led Abu Dhabi Dialogue. In 2006, The Abu Dhabi Dialogue (ADD) was created to foster cooperation in the management of the rivers in the Greater Himalayas (SAWI 2015c). The ADD process was not limited only to the Ganges–Brahmaputra problemshed, but rather included all seven countries (Afghanistan, Bangladesh, Bhutan, China, India, Nepal and Pakistan) that share the rivers of the Himalayas. The members of the ADD were not only government officials, but also representatives from academia and civil society. In other words, the ADD was a formalised Track II dialogue process. It is also modelled on the common water-sharing framework in line with the Nile Basin Initiative (Earle et al. 2015, p. 41).

According to the World Bank, '[t]he key rationale for engagement is to demonstrate and then to help achieve the mutual benefits of cooperation across shared river basins' (World Bank 2013). The focus of the ADD and SAWI on fostering regional cooperation over transboundary water issues is an appropriate one considering the myriad unresolved transboundary water issues in the

Ganges–Brahmaputra problemshed. The need for increasing transboundary water cooperation has been noted by scholars from within the region, yet the riparian states themselves are on the whole unwilling or unable to actively pursue positive water interaction with one another. The riparians are also largely indifferent to the efforts of international actors to facilitating cooperative processes. What hope, then, do these third parties have of improving transboundary water interactions in the Ganges–Brahmaputra problemshed?

How can international actors improve transboundary water governance in developing countries?

Foreign aid donors and international organisations have taken on the challenge of increasing transboundary water cooperation in the Ganges–Brahmaputra problemshed, but they are unlikely to meet it singlehandedly. As Earle et al. (2015, p. 41) argue, the attempts to improve transboundary water interaction 'have not ventured beyond political statements and setting up lofty goals' and 'the World Bank has failed in its hope to bring South Asian countries within a common water-sharing framework'.

There are many explanations for this, but three in particular reveal something about the nature of transboundary water cooperation. First, riparian states seem unwilling to address unresolved transboundary water issues. Earle et al. (2015, pp. 41–42) note that 'there is certainly a lack of political will in the basin, particularly in India, to provide the space for multilateral river institutions to effectively emerge'.

Desire for cooperation is important because, in the words of Israeli hydrologist Uri Shamir (cited in Julien 2012, p. 62), '[i]f there is a political will for peace, water will not be a hindrance. If you want reasons to fight, water will give you ample opportunities'. India, Nepal and Bangladesh (but not Bhutan) do not place transboundary water governance at the top of their political priorities, and ongoing tensions over past failures at improving interaction continue to keep political will low.

The main argument here is that international actors may be able to play a role – albeit a highly circumscribed one – in creating the conditions necessary for transboundary cooperation to emerge, if they are invited to do so. Of course, if there is trust between riparians, and political will to address transboundary water issues, then cooperative water interaction will likely develop without the assistance of third parties. International actors can, however, play a role in facilitating Track II dialogues and these dialogues may in the long term contribute to the development of transboundary water cooperation. But, as Jones (2015, p. 140) points out, 'real peace is made as a result of a complex and interlocking web of factors, and … Track Two, however important in getting a dialogue going, is but one of these.' Similarly, Pahl-Wostl (2015, p. 32) shows that '[g]iven the complex nature of governance systems, change can be expected to be a combination of purposeful collective action and emergent phenomena resulting from selforganising processes and the interactions among a range of actors'.

System-wide change takes a long time. Long-term engagement is, according to Pohl et al. (2014, p. 35), the only appropriate approach to successfully building trust in a system of transboundary water interaction. But even in the long term cooperation cannot be guaranteed. There is no one path to cooperation, and no certainty for the external actors that they will resolve conflicts. Numerous factors play into the dynamics within dispute resolution processes and the contexts in which they occur: culture, domestic politics, trust (or lack thereof) between parties, power asymmetries, issue power, the presence or absence of third parties and so on. While all these, and other, factors define the outcome of interactions, dialogues and negotiations, they do not pre-determine them. As Hill (2015, p. 732) shows, there have in the past been many initiatives in the Ganges–Brahmaputra problemshed for water dispute resolution, but changing the way riparian states think about shared water resources is difficult: 'Given how politically contentious many parts of the region already are, this will be a long and unenviably difficult process and there are no guarantees of any success in this regard.' But international actors can improve what they do.

International organisations and foreign aid donors may be better placed to support riparian states in influencing the conditions in which transboundary water cooperation may emerge if they collaborate more closely with stakeholders within the basin and address their concerns. Specifically, stakeholders within the Ganges–Brahmaputra problemshed would like to see approaches to transboundary water interaction that:

- address the political context and historical grievances;
- build trust;
- address power asymmetry between riparian states;
- create political will for cooperation;
- focus on water justice;
- de-securitise water;
- take a problemshed view;
- strengthen water sharing institutions; and
- move beyond narratives of water scarcity and supply-side solutions.

Those within the Ganges–Brahmaputra would generally like to see third parties better tackle the historical grievances and root causes of water conflicts – even (or especially) if those causes are not related to water directly. They would also like to see transboundary waters de-securitised, and water sharing institutions strengthened. At the same time, the power asymmetry between riparian states must be mitigated. These approaches may help build trust between riparian states, and create political will for addressing shared water problems. Of course these approaches are not intended as a one-size-fits-all solution to complex challenges. Indeed, it is not likely that there will ever be a single solution or best approach given the interconnectedness and complexity of water governance, food security, climate change, sustainable development and so on.

Nonetheless, what international actors *are* currently doing in this problemshed is not always leading to the desired outcomes. Have international organisation and foreign aid donors directly or substantially contributed to improved transboundary water interaction in the Ganges–Brahmaputra problemshed? Earle et al. (2015, p. 41) argue that no, unfortunately, international actors have not yet been able to bring South Asian states within a common water-sharing framework. There are several explanations for this. There are limitations inherent in the way that international actors conduct their interventions into transboundary water governance. As one analyst put it, there are 'a lot of solutions looking for problems' in the Ganges–Brahmaputra problemshed, and 'when all you have is a hammer, then every problem becomes a nail' (Anonymous 10, 2014). Thus, Track II dialogue, knowledge-sharing, basin-wide approaches and so on miss the mark because they are general tactics, not tailored to the specific context of the Ganges–Brahmaputra problemshed.

There are also structural limitations on international actors. The system in which international organisations and donor agencies operate – the system of foreign aid – is crowded, inefficient, short-sighted and driven by complex agendas. Furthermore, the effectiveness and impact of international actors in the transboundary water governance of the Ganges–Brahmaputra problemshed will likely never be known because of the twin problems of effect and attribution (Mayne 1999, pp. 2–3). The crowded and complex system of foreign aid means that it is almost impossible to separate out the effects of individual donors and external factors on common objectives (Riddell 2014, p. 26). And the desired effects can be brought about by factors entirely unrelated to the project or programme working towards them (Riddell 2014, p. 45).

It is unlikely that the third parties engaged in improving transboundary water interaction in the Ganges–Brahmaputra problemshed will achieve in the foreseeable future their stated goals (listed in Chapter 6) of increasing cooperation, instituting good water governance, ensuring overall regional stability, building climate change resilience and sustainable development. But how much does that matter, and why?

As discussed in Chapter 11, proving the success of development projects and programmes is used to 'justify their very existence' (Riddell 2014, p. 1). Yet, cause and effect is rarely, if ever, direct and measurable in a complex adaptive system such as hydropolitics. To some extent, then, the difficulty of creating transboundary water cooperation is not due to inefficient or incapable third parties, but rather to the complexity and unpredictability of the system.

Transboundary water interaction can be understood as an ambiguous, largely unquantifiable, and nebulous process that arises out of largely unpredictable dynamics in a complex system of other political, social, economic and environmental dynamics that interplay on local, national, regional and global scales. In the Ganges–Brahmaputra problemshed, foreign-led interventions are only one component of transboundary water interaction. Improving the quality of interaction is unlikely to occur through the efforts of one actor, stakeholder group or initiative. Instead, transboundary water cooperation involves slowly but steadily 'chipping away' at aspects of the problem by numerous actors working in concert over a long period of time, and it can never be achieved in one foreign aid funding cycle or adequately measured by the sort of short-term indicators favoured by foreign aid donors (Prasai 2014). Foreign-led interventions in the transboundary water governance of the Ganges–Brahmaputra problemshed are merely a drop in the ocean in terms of their overall effect on water cooperation. Yet, the ocean is comprised of individual drops and every one counts.

Glossary

Basin: See: River basin

Benefit sharing: Any action designed to change the allocation of costs and benefits associated with cooperation.

Co-riparians: Two or more countries that share a river.

Deliberative water governance: Constructive engagement in water governance arenas through promotion of inclusive, deliberative processes that emphasise different perspectives, critical analysis, learning and institution-building whilst respecting rights, accounting for risks, acknowledging responsibilities and fairly distributing rewards.

Distributive water justice: The allocation of water resources that is perceived as fair by stakeholders. Principles of distributive water justice include: equitable sharing of benefits; avoiding unfair and involuntary risk-bearing; protection of livelihood security; and provision of compensation, insurance and, where necessary, welfare support for project-affected people. See also: Water justice, Procedural water justice

Energy security: The ability of a nation to secure sustainable energy supplies to meet its energy needs at a reasonable price.

Equitable distribution of water resources: The term 'equality' is used when discussing the assumed parity of capacity and rights of actors, and 'equitability' when discussing outcomes that may be considered balanced and fair, though not necessarily equal (so, in the same sense that others use the term 'equity'). 'Equitability' can be considered the spirit of justice. See also: Water justice

Good governance: Good governance is participatory, consensus orientated, accountable, transparent, responsive, equitable and inclusive, effective and efficient, and follows the rule of law.

Hydro sovereignty: The principle that riparian states have exclusive rights over water flowing in their territory.

Hydro-diplomacy: Diplomacy, sometimes facilitated by third parties, to anticipate, mitigate and resolve conflicting claims on shared water resources.

Hydro-hegemon: A riparian state that can establish the form of interaction over transboundary waters that it prefers. Whether the hydro-hegemon chooses leadership or domination in the water sector is determined by the broader political context.

Hydropolitics: The political dynamics that determine who decides who gets water, when, how and why.

Integrated Water Resources Management (IWRM): A process which promotes the coordinated development and management of water, land and related resources, in order to maximise the resultant economic and social welfare in an equitable manner without compromising the sustainability of vital ecosystems.

Internally renewable water resources: Amount of runoff generated within a country, plus any flows into the country, less any flows to a neighbouring country. See also: Water dependency

International watercourse: A system of surface waters and ground waters constituting by virtue of their physical relationship a unitary whole and normally flowing into a common terminus, parts of which are situated in different states.

Necessary condition: A condition that must be satisfied for a statement to be true, but that does not in and of itself make it true. See also: Sufficient condition

Negative transboundary water interaction: Arrangements that do not satisfy the interests of at least one of the parties involved, thus directly or indirectly degrading relations at the broader political level. See also: Positive transboundary water interaction

Negotiation: Formal, bounded conflict usually over a single issue. Negotiation is a joint decision-making process through which negotiating parties accommodate their conflicting interests into a mutually acceptable settlement. A negotiation can be in aid of dispute resolution or for the purpose of reaching a binding agreement or treaty. A negotiation can involve a third party in the role of mediator, conciliator, arbitrator or adjudicator. Negotiation is the means of dispute resolution most often employed by states when trying to resolve any international conflict, including those over transboundary water resources. See also: Pre-negotiation

Non-zero-sum thinking: A joint problem-solving framework that assumes outcomes favourable to one party are not necessarily detrimental to another. See also: Zero-sum thinking

Positive-sum thinking: See: Non-zero-sum thinking

Positive transboundary water interaction: Contexts where the interaction satisfies the interests of all players and culminates in an improvement (or at least a sustenance) of relations at the broader political level. See also: Negative transboundary water interaction

Pre-negotiation: Steps that take place prior to negotiations in an effort to ensure the best 'groundwork' is in place for effective negotiation. Includes identifying the right stakeholders, developing an agenda and timetable for the negotiation, and establishing joint problem-solving procedures. Usually refers to the work that individual parties do on their own to prepare for negotiations as well as the work done on behalf of all the participants by a convener or a professional neutral. See also: Negotiation

Problemshed: A geographic area that is large enough to encompass water management issues, but small enough to make implementation feasible.

Procedural water justice: The process of allocating water resources that is perceived as fair by the stakeholders. Principles of procedural water justice include: inclusion of stakeholders in decision-making; access to information, legal and other support; demonstrable public acceptance; and free, prior and informed consent of affected indigenous and tribal peoples. See also: Water justice, Distributive water justice

Riparian: An area near or on the banks of a river or other major body of water. Also referred to in the context of 'upper' and 'lower' riparians, which designate the country in which a certain transboundary river originates.

Riparian state: The country or other political entity on or near the banks of a river or other major body of water.

River basin: The area drained by a river and its tributaries.

Securitisation: Elevating a non-security issues to a security issue. Securitising an environmental issue implies subjecting it to a military security logic (e.g., top-down decision-making, classifying information, bypassing normal standards of participation and economic valuation.)

Sufficient condition: A condition that must be satisfied for a statement to be true and without which the statement cannot be true. See also: Necessary condition

Sustainable development: Development that meets the needs of the present without compromising the ability of future generations to meet their own needs.

Third party: A person or group besides the two primarily involved in a situation, especially a dispute. A third party is often a feature of Track II dialogue. Track II dialogue features two parties who are in conflict plus a third party who is present to arrange and facilitate the interaction. Not all Track II processes require an outside third party, but the great majority of conflict resolution-orientated Track II processes feature a third party. In the context of transboundary water interaction in the Ganges–Brahmaputra problemshed, it is any individual, government, organisation or other group that does not come from India, Nepal, Bhutan or Bangladesh but that nonetheless is engaged in water governance. See also: Track II dialogue

Track II dialogue: Unofficial dialogues, generally between two antagonistic parties, and often facilitated by an impartial third party and involving individuals with some close connections to their respective official communities, focused on cooperative efforts to explore new ways to resolve differences over, or discuss new approaches to, policy-relevant issues.

Transboundary water cooperation: Emergent property of positive transboundary water interaction. See also: Transboundary water interaction, Positive transboundary water interaction

Transboundary water interaction: Relations of co-existing cooperation and conflicts among communities, groups or states over international or sub-national water, with a focus here on inter-state interaction. See also: Positive transboundary water interaction, Negative transboundary water interaction

Transboundary waters: Any surface or groundwaters which mark, cross or are located on boundaries between two or more states. Wherever transboundary waters flow directly into the sea, these transboundary waters end at a straight line across their respective mouths between points on the low-water line of their banks.

Water dependency: The percentage of water that originates outside a country's political borders. See also: Internally renewable water resources

Water governance: The range of political, social, economic and administrative systems that are in place to regulate development and management of water resources and provisions of water services at different levels of society. See also: Deliberative water governance

Water justice: Water governance that aims for achieving equitable distribution of water resources. Distributive and procedural justice are the most popular means of understanding justice because they provide good insights into the quantities and processes of allocation and distribution of natural resources. In complex multi-level systems just and unjust outcomes can result from the same water allocation decision. In some cases, a just outcome at one level may cause an injustice at another level for the same or a different set of stakeholders. What appears to be just at one level may not be just at other levels within the system under investigation. A sense of unfairness and injustice is often at the root of conflict, over water or otherwise. See also: Procedural water justice, Distributive water justice

Water scarcity: The point at which the aggregate impact of all users impinges on the supply or quality of water under prevailing institutional arrangements to the extent that the demand by all sectors, including the environment, cannot be fully satisfied.

Water security: The capacity of a population to safeguard sustainable access to adequate quantities of acceptable quality water for sustaining livelihoods, human well-being and socio-economic development, for ensuring protection against water-borne pollution and water-related disasters, and for preserving ecosystems in a climate of peace and political stability.

Water stress: The volume of freshwater resources in a country divided by the population. Countries with less than 1,666 cubic metres of water availability per capita are considered water stressed, while those with less than 1,000 cubic metres per capita are chronically water stressed.

Watershed: See: River basin

Zero-sum thinking: A win/lose perspective on negotiations that assumes gains to one party result in matching losses to other parties. In water negotiations, this thinking manifests itself in the notion that water can only be used once, and by only one party. See also: Non-zero-sum thinking

References

VI BRICS Summit 2014, Agreement on the New Development Bank – Fortaleza, 15 July, viewed 7 May 2015, http://brics6.itamaraty.gov.br/media2/press-releases/219-agreement-on-the-new-development-bank-fortaleza-july-15

Agnew, C & Woodhouse, P 2011, *Water Resources and Development*, Routledge, New York.

Allan, JA 2006, 'IWRM: The New Sanctioned Discourse?' in P Mollinga, A Dixit & K Athukorala (eds), *Integrated Water Resources Management: Global Theory, Emerging Practice, and Local Needs*, Sage Publications, New Delhi, pp. 38–63.

Anonymous 1 2014, personal communication, 19 May 2014, New Delhi.

Anonymous 3 2014, personal communication, 16 October 2014, Dhaka.

Anonymous 4 2014, personal communication, 18 July 2014, Mumbai.

Anonymous 5 2014, personal communication, 19 August 2014, Kathmandu.

Anonymous 6 2014, personal communication, 26 June 2014, New Delhi.

Anonymous 7 2014, personal communication, 7 August 2014, Kathmandu.

Anonymous 9 2014, personal communication, 14 October 2014, Dhaka.

Anonymous 10 2014, personal communication, 2 November 2014, New Delhi.

Anonymous 11 2014, personal communication, 23 May 2014, New Delhi.

Araral, E & Wang, Y 2013, 'Water Governance 2.0: A Review and Second Generation Research Agenda', *Water Resources Management*, vol. 27, pp. 3945–3957.

Asthana, V & Shukla, AC 2014, *Water Security in India; Hope, Despair, and the Challenges of Human Development*, Bloomsbury Academic, New York.

Attwood, DW 2007, 'Small is Deadly, Big is Wasteful: The Impact of Large-scale Industrial Systems in Western India' in A Baviskar (ed.), *Waterscapes: The Cultural Politics of a Natural Resource*, Permanent Black, New Delhi, pp. 11–39.

Australian Agency for International Development 2013, *South Asia Regional Development Program Strategy 2013–2017*, July, AusAID, Canberra.

Australian Water Partnership 2016, 'Overview', Australian Water Partnership, viewed 14 April 2016, waterpartnership.org.au/about/#page-12018

Bandyopadhyay, J, Mallik, B, Mandal, M & Perveen, S 2002, 'Dams and Development: Report on a Policy Dialogue', *Economic and Political Weekly*, vol. 37, no. 40, pp. 4108–12.

Banki, S 2014, 'Finding a Future for Minorities in Bhutan's Emerging Democracy', *South East Asia Forum*, 29 May, viewed 8 June 2014, www.eastasiaforum.org/2014/05/29/finding-a-future-for-minorities-in-bhutans-emerging-democracy/

Baqai, H 2011, *Non-Traditional Sources of Conflict in South Asia 1971–2000*, VDM Verlag, Saarbruecken.

Baral, N & Heinen, JT 2005, 'The Maoist People's War and Conservation in Nepal', *Politics and the Life Sciences*, vol. 24, nos. 1/2, pp. 2–11.

Baviskar, A (ed.) 2007, *Waterscapes: The Cultural Politics of a Natural Resource*, Permanent Black, New Delhi.

Behera, NC 2002, 'Forging New Solidarities: Nonofficial Dialogues' in M Mekenkamp, P van Tongeren & H van de Veen (eds), *Searching for Peace in Central and South Asia: An Overview of Conflict Prevention and Peacebuilding Activities*, Lynne Rienner Publishers, Boulder, pp. 210–236.

Bhaduri, A & Kaushal, S 2014, personal communication, 25 July 2014, New Delhi.

Bharadwaj, A, Tongia, R & Arunachalam, VS 2006, 'Wither Nuclear Power?', *Economic and Political Weekly*, vol. 41, no. 12, pp. 1203–1212.

Bhasin, M 2008, 'India's Role in South Asia – Perceived Hegemony or Reluctant Leadership?', *Indian Foreign Affairs Journal*, vol. 3, no. 4.

Bhattarai, R 2005, *Geopolitics of Nepal and International Responses to Conflict Transformation*, Friends for Peace, FFP Publications Series 006, Kathmandu.

Bhonsale, M 2014, personal communication, 20 October 2014, Kolkata.

Bhushal, R 2016, 'Will China Invest in Nepal's Hydropower Plants? Nepal Hopes to Finally Secure Chinese Investment on Major Hydropower Dams During the Prime Minister's Trip to Beijing Next Week, as well as Key Energy Agreements', *The Third Pole*, 17 March, viewed 2 May 2016, www.thethirdpole.net/2016/03/17/will-china-invest-in-nepals-hydropower-plants/

Biswas, AK 2001, 'Management of International Rivers: Opportunities and Constraints' in AK Biswas & JI Uitto (eds), *Sustainable Development of the Ganges–Brahmaputra–Meghna Basins*, Oxford University Press, New Delhi, pp. 1–16.

Biswas, AK & Uitto, JI (eds) 2001, *Sustainable Development of the Ganges–Brahmaputra–Meghna Basins*, Oxford University Press, New Delhi.

Brant, P 2014, 'Chinese Foreign Aid', Lowy Institute for International Policy, viewed 7 May 2015, www.lowyinstitute.org/issues/chinese-foreign-aid

Briscoe, J 2010, 'Overreach and Response: The Politics of the WCD and its Aftermath', *Water Alternatives*, vol. 3 no. 2, pp. 399–415.

Bygstad, B & Munkvold, BE 2011, 'In Search of Mechanisms: Conducting a Critical Realist Data Analysis', completed research paper, Thirty-Second International Conference on Information Systems, Shanghai 2011.

Campbell, KM, Gulledge, J, McNeill, JR, Podesta, J, Ogden, P, Fuerth, L, Woolsey, RJ, Lennon, ATJ, Smith, J, Weitz, R & Mix, D 2007, *The Age of Consequences: The Foreign Policy and National Security Implications of Global Climate Change*, Center for Strategic and International Studies, Center for a New American Security, Washington DC.

Cascão, AE & Zeitoun, M 2010, 'Power, Hegemony and Critical Hydropolitics' in A Earle, A Jägerskog & J Öjendal (eds), *Transboundary Water Management: Principles and Practice*, Earthscan, London, pp. 27–42.

Chapman, C 2014, 'Australia Now Expected to Join Asian Infrastructure Investment Bank', Australian Institute of International Affairs, viewed 7 May 2015, www.internationalaffairs.org.au/australian_outlook/australia-now-expected-to-join-asian-infrastructure-investment-bank/

Chellaney, B 2011, *Water: Asia's New Battleground*, Georgetown University Press, Washington DC.

Chellaney, B 2013, *Water, Peace and War: Confronting the Global Water Crisis*, Rowman & Littlefield Publishers, Lanham.

Chellaney, B 2014, 'Water, Power, and Competition in Asia', *Asian Survey*, vol. 54, no. 4, pp. 621–650.

Chokkakula, S 2012, Disputes, (de)Politicization and Democracy: Interstate Water Disputes in India, Working Paper no. 108, RULNR Working Paper no. 13, January, Centre for Economic and Social Studies, Research Unit for Livelihoods and Natural Resources, Hyderabad.

CIA World Factbook, Bhutan 2015, viewed 25 June 2015, www.cia.gov/library/publications/the-world-factbook/geos/bt.html

Cleaver, F & de Koning, J 2015, 'Furthering Critical Institutionalism', *International Journal of the Commons*, vol. 9, no. 1, pp. 1–18.

Colopy, C 2012, *Dirty, Sacred Rivers: Confronting South Asia's Water Crisis*, Oxford University Press, New York.

Conca, K 2006, 'The New Face of Water Conflict' in *Navigating Peace*, no. 3, November, Woodrow Wilson International Centre for Scholars, Environmental Change and Security Program, Washington DC.

Condon, E, Hillman, P, King, J, Lang, K & Patz, A 2009, 'Resource Disputes in South Asia: Water Scarcity and the Potential for Interstate Conflict', prepared for the Office of South Asia Analysis, US Central Intelligence Agency, Workshop in International Public Affairs, 1 June, Robert M La Follette School of Public Affairs, University of Wisconsin-Madison.

Collins, A 2007, *Contemporary Security Studies*, Oxford University Press, Bath.

Cosgrove, WJ 2003, *Water Security and Peace: A Synthesis of Studies Prepared under the PCCP-Water for Peace Process*, UNESCO, Technical Documents in Hydrology, PC→CP series, no. 29, SC-2003/WS/51, UNESCO, Paris.

Cullet, P 2007, 'Water Law in India: Overview of Existing Framework and Proposed Reforms', Working Paper 2007-01, International Environmental Law Research Centre, Geneva.

Dash, KC 2008, *Regionalism in South Asia: Negotiating Cooperation, Institutional Structures*, Routledge, New York.

Davies, R & Evans, JW 2015, 'Conclusion' in JW Evans & R Davies (eds), *Too Global to Fail: The World Bank at the Intersection of National and Global Public Policy in 2025*, Directions in Development, doi:10.1596/978-1-4648-0307-9, World Bank, Washington DC, pp. 265–276.

Department for International Development, Government of the United Kingdom (DFAT) n.d., 'India Country Brief', viewed 25 May 2015, http://dfat.gov.au/geo/india/Pages/india-country-brief.aspx

Department for International Development, Government of the United Kingdom (DFAT) 2012, 'South Asia Water Governance Programme (SAWGP) Business Case and Summary', GB-1-202775, October, DFID, London, https://devtracker.dfid.gov.uk/projects/GB-1-202775/documents

Department of Foreign Affairs and Trade, Government of Australia (DFAT) 2014, *Aid Program Performance Report 2013–14: South Asia Regional Program*, September, DFAT, Canberra.

Devex 2014, 'In Latest Indian Budget, Aid Spending Dwarfs Aid Receipts', 24 February, viewed 2 May 2015, www.devex.com/news/in-latest-indian-budget-aid-spending-dwarfs-aid-receipts-82915

Dhaka, A 2005, *South Asia and Central Asia: Geopolitical Dynamics*, Mangal Deep Publications, Jaipur.

Dinar, A, Dinar, S, McCaffrey, S & McKinney, D 2007, *Bridges Over Water: Understanding Transboundary Water Conflict, Negotiation and Cooperation*, World Scientific Series on Energy and Resource Economics – vol. 3, World Scientific Publishing Co, Singapore.

Dinar, S 2002, 'Water, Security, Conflict and Cooperation', *SAIS Review*, vol. 22, no. 2, pp. 229–253.

Dixit, A & Gyawali, D 2010, 'Nepal's Constructive Dialogue on Dams and Development', *Water Alternatives*, vol. 3 no. 2, pp. 106–123.

Dore, J 2007, 'Mekong Region Water-Related MSPs: Unfulfilled Potential' in J Warner (ed.), *Multi-Stakeholder Platforms for Integrated Water Management*, Routledge, Oxford, pp. 205–234.

Dore, J 2014, 'An Agenda for Deliberative Water Governance Arenas in the Mekong', *Water Policy*, vol. 16, pp. 194–214.

D'Souza, R 2007, 'From Calamity to Resource: Flood Control and the Politics of Natural Limits' in A Baviskar (ed.), *Waterscapes: The Cultural Politics of a Natural Resource*, Permanent Black, New Delhi, pp. 248–280.

D'Souza, R 2012, 'Hydropolitics, the Indus Water Treaty and Climate Change: Writing a New Script for the Indus Rivers' in L Powell & S Mittra (eds), *Perspectives on Water: Constructing Alternative Narratives*, Academic Foundation, New Delhi, pp. 25–38.

D'Souza, R., 2014, personal communication, 23 June 2014, New Delhi.

Earle, A, Cascão, AE, Hansson, S, Jägerskog, A, Swain, A & Öjendal, J 2015, *Transboundary Water Management and the Climate Change Debate*, Earthscan Studies in Water Resource Management, Routledge, Abingdon.

Elver, H 2006, 'International Environmental Law, Water and the Future', *Third World Quarterly*, Reshaping Justice: International Law and the Third World, vol. 27, no. 5, pp. 885–901.

Emmers, R 2007, 'Securitization' in A Collins (ed.), *Contemporary Security Studies*, Oxford University Press, Bath, pp. 109–125.

Evans, A 2010, 'Resource Scarcity, Climate Change and the Risk of Violent Conflict', World Development Report 2011, Background Paper, 9 September, Centre on International Cooperation, New York University, New York.

Evans, JW & Davies, R (eds) 2015 *Too Global to Fail: The World Bank at the Intersection of National and Global Public Policy in 2025*, Directions in Development, doi:10.1596/978-1-4648-0307-9, World Bank, Washington DC.

Evans, JW & Davies, R 2015, 'Introduction' in JW Evans & R Davies (eds), *Too Global to Fail: The World Bank at the Intersection of National and Global Public Policy in 2025*, Directions in Development, doi:10.1596/978-1-4648-0307-9, World Bank, Washington DC, pp. 1–10.

Evans, JW 2015a, 'Environmental Sustainability as a Development Issue: The Evolution of an Idea' in JW Evans & R Davies (eds), *Too Global to Fail: The World Bank at the Intersection of National and Global Public Policy in 2025*, Directions in Development, doi:10.1596/978-1-4648-0307-9, World Bank, Washington DC, pp. 53–66.

Evans, JW 2015b, 'The Future Is Now: Scenarios to 2025 and Beyond' in JW Evans & R Davies (eds), *Too Global to Fail: The World Bank at the Intersection of National and Global Public Policy in 2025*, Directions in Development, doi:10.1596/978-1-4648-0307-9, World Bank, Washington DC, pp. 67–92.

Evans, JW, Nicholson, I, TuiSamoa LeLaulu, L & Wescott, A 2015, 'Shifting Priorities: Re-envisioning World Bank Partnerships for Transformational Impact' in JW Evans & R Davies (eds), *Too Global to Fail: The World Bank at the Intersection of National and*

Global Public Policy in 2025, Directions in Development, doi:10.1596/978-1-4648-0307-9, World Bank, Washington DC, pp. 189–204.
eWater 2012, 'New Australia–India Partnership', 19 October, viewed 26 March 2015, www.ewater.com.au/news/media/?news=214
Falkenmark, M, Berntell, A, Jägerskog, A, Lundqvist, J, Matz, M & Tropp, H 2007, 'On the Verge of a New Water Scarcity: A Call for Good Governance and Human Ingenuity', Stockholm International Water Institute, Stockholm.
Faure, GO & Rubin, JZ (eds) 1993, *Culture and Negotiation: The Resolution of Water Disputes*, Sage, Newbury Park, CA.
Faure, GO & Sjöstedt, G 1993, 'Culture and Negotiation: An Introduction' in GO Faure & JZ Rubin (eds), *Culture and Negotiation: The Resolution of Water Disputes*, Sage, Newbury Park, CA, pp. 1–13.
Feldman, DL 2007, *Water Policy for Sustainable Development*, The Johns Hopkins University Press, Baltimore.
Garrick, D 2014, 'Systemic Risks and Fragmented Responses: The Challenge of the Colorado River' Asia and the Pacific Policy Society, Policy Forum, viewed 8 December 2014, www.policyforum.net/systematic-risks-fragmented-responses/#sthash.3R0Ufe8N.dpuf
Ghosh, D, Goodall, H & Hemelryk-Donald, S (eds) 2009, *Water, Sovereignty and Borders in Asia and Oceania*, Routledge, New York.
Global Water Partnership 2003, *Effective Water Governance: Learning from the Dialogues*, GWP Secretariat, Stockholm.
Goodland, R 2010, 'The World Bank Versus the World Commission on Dams', *Water Alternatives*, vol. 3 no. 2, pp. 384–398.
Government of Australia 2012, *Australia in the Asian Century*, White Paper, October 2012, Government of Australia, Canberra.
Government of India 1987, National Water Policy (1987), Ministry of Water Resources, Government of India, New Delhi.
Government of India 2002, National Water Policy (2002), Ministry of Water Resources, Government of India, New Delhi.
Government of India 2012, Draft National Water Policy (2012) as Recommended by National Water Board in its 14th Meeting Held on 7th June, 2012, Ministry of Water Resources, Government of India, New Delhi.
Government of India 2012, National Water Policy (2012), Ministry of Water Resources, Government of India, New Delhi.
Government of India 2015, 'River Development & Ganga Rejuvenation', press release 17 January, Ministry of Water Resources, viewed 16 May 2015, www.indiawaterweek.in/docs/2015/press_release.pdf
Government of Norway 2011, *Towards Greener Development: On a Coherent Environmental and Development Policy*, Meld.St. 14 Report to the Storting (white paper) Recommendation from the Ministry of Foreign Affairs of 8 April 2011, approved in the Council of State the same date, (White paper from the Stoltenberg II Government), Government of Norway, Oslo.
Granit, J & Claassen, M 2009, 'A Path Towards Realising Tangible Benefits in Transboundary River Basins' in A Jägerskog & M Zeitoun (eds), *Getting Transboundary Water Right: Theory and Practice for Effective Cooperation*, Report no. 25, Stockholm International Water Institute, Stockholm, pp. 21–26.
Grey, D, Sadoff, C & Connors, G 2009, 'Effective Cooperation on Transboundary Waters: A Practical Perspective' in A Jägerskog & M Zeitoun (eds), *Getting*

Transboundary Water Right: Theory and Practice for Effective Cooperation, Report no. 25, Stockholm International Water Institute, Stockholm, pp. 15–20.
Groenfeldt, D 2010, 'Viewpoint – The Next Nexus? Environmental Ethics, Water Policies, and Climate Change', *Water Alternatives* vol. 3 no. 3, pp. 575–586.
Gupta, J., 2014, personal communication, 1 July 2014, New Delhi.
Gyawali, D 2001, 'Missing Leg: South Asia's Hobbled Water Technology Choices', *Economic and Political Weekly*, vol. 36, no. 39, pp. 3743–3758.
Gyawali, D 2013, 'Bringing Hydro-Diplomacy into Nepal's Economic Diplomacy' in M Shrestha & A Shakya (eds), *Vision on Foreign Affairs*, International Concern Center, Kathmandu, pp. 197–222.
Gyawali, D & Dixit, A 1999, 'Mahakali Impasse and Indo-Nepal Water Conflict', *Economic and Political Weekly*, vol. 34, no. 9, pp. 553–564.
Habeeb, WM 1988, *Power and Tactics in International Negotiation: How Weak Nations Bargain with Strong Nations*, The Johns Hopkins University Press, Baltimore.
Hanasz, P 2014a, 'The Problem with Problems of Water Scarcity in South Asia', GWF Discussion Paper 1408, February, Global Water Forum, Canberra, viewed 2 October 2016, www.globalwaterforum.org/2014/02/25/the-problem-with-problems-of-water-scarcity-in-south-asia/
Hanasz, P 2014b, 'Sharing Waters vs Sharing Rivers: The 1996 Ganges Treaty', 28 July, Global Water Forum, Canberra, viewed 2 October 2016, www.globalwaterforum.org/2014/07/28/sharing-waters-vs-sharing-rivers-the-1996-ganges-treaty/
Hanasz, P 2015a, 'Blue Gold for Whom? Multi-Level Games in the Development of Himalayan Hydropower', International Journal of Water Governance – Special Issue 2015, doi:10.7564/13-IJWG32, pp. 1–18.
Hanasz, P 2015b, 'The Politics of Water Governance in the Ganges–Brahmaputra–Meghna Basin', ORF Issue Brief, November, Issue no. 112, Observer Research Foundation, New Delhi.
Hangzo, PKK 2012, 'Transboundaray Rivers in the Hindu-Kush-Himalaya (HKH) Region: Beyond the "Water as Weapon" Rhetoric', Insight NTS Insight September, Centre for Non-Traditional Security Studies, S. Rajaratnam School of International Studies, Singapore.
Hardiman, D 2008, 'The Politics of Water in Colonial India: The Emergence of Control' in K Lahiri-Dutt & RJ Wasson (eds), *Water First: Issues and Challenges for Nations and Communities in South Asia*, Sage, New Delhi, pp. 47–58.
Harris, LM, Goldin, JA & Sneddon, C 2013, *Contemporary Water Governance in the Global South; Scarcity, Marketization and Participation*, Routledge, Abingdon.
Hill, D 2008, 'The Regional Politics of Water Sharing: Contemporary Issues in South Asia' in K Lahiri-Dutt & RJ Wasson (eds), *Water First: Issues and Challenges for Nations and Communities in South Asia*, Sage, New Delhi, pp. 59–80.
Hill, D 2009, 'Boundaries, Scale and Power in South Asia' in D Ghosh, H Goodall & S Hemelryk-Donald (eds), *Water, Sovereignty and Borders in Asia and Oceania*, Routledge, New York, pp. 87–103.
Hill, D 2013, 'Trans-Boundary Water Resources and Uneven Development: Crisis Within and Beyond Contemporary India', *South Asia: Journal of South Asian Studies*, vol. 36 no. 2, pp. 243–257.
Hill, D 2015, 'Where Hawks Dwell on Water and Bankers Build Power Poles: Transboundary Water, Environmental Security and the Frontiers of Neo-liberalism', *Strategic Analysis*, vol. 39 no. 6, doi:10.1080/09700161.2015.1090679, pp. 729–743.

Homer-Dixon, T 1995, 'The Ingenuity Gap: Can Poor Countries Adapt to Resource Scarcity?', *Population and Development Review*, the Population Council, vol. 21, no. 3, pp. 587–612.

Homer-Dixon, T 1998, 'Environmental Scarcity and Intergroup Conflict' in M Klare & Y Chandrani (eds), *World Security: Challenges for a New Century*, third edition, St Martins Press, New York, pp. 342–365.

Hydro World 2015, 'CTG Receives Final Approvals for Nepal's 750-MW West Seti Hydropower Project', 15 April, viewed 10 July 2017, www.hydroworld.com/articles/2015/04/ctg-receives-final-approvals-for-nepal-s-750-mw-west-seti-hydropower-project.html

International Centre for Integrated Mountain Development 2016, 'Strategic Framework and Goals', viewed 14 April 2016, www.icimod.org/?q=5

Islam, S & Susskind, LE 2013, *Water Diplomacy: A Negotiated Approach to Managing Complex Water Networks*, RFF Press, New York.

Iyer, RR 2001, 'World Commission on Dams and India: Analysis of a Relationship', *Economic and Political Weekly*, vol. 36, no. 25, pp. 2275–2281.

Iyer, RR 2007, *Towards Water Wisdom: Limits, Justice, Harmony*, Sage, New Delhi.

Iyer, RR 2008a, 'Floods, Himalayan Rivers, Nepal: Some Heresies', *Economic and Political Weekly*, vol. 43, no. 46, pp. 37–40.

Iyer, RR 2008b, 'National and Regional Water Concerns: Setting the Scene' in K Lahiri-Dutt & RJ Wasson (eds), *Water First: Issues and Challenges for Nations and Communities in South Asia*, Sage, New Delhi, pp. 3–30.

Iyer, RR 2010, 'Resolving River Water Disputes in India: Reflections' in NS Mohan, S Routray & N Sashikumar (eds), *River Water Sharing: Transboundary Conflict and Cooperation in India*, Routledge, New Delhi, pp. 66–80.

Jägerskog, A & Zeitoun, M 2009, *Getting Transboundary Water Right: Theory and Practice for Effective Cooperation*, Report no. 25, Stockholm International Water Institute, Stockholm.

Jägerskog, A, Swain, A & Öjendal, J 2014, *Water Security Volume III: Water Security and Development – An Intimate Relation*, Sage, London.

Jain, BM 2010, *India in the New South Asia: Strategic, Military and Economic Concerns in the Age of Nuclear Diplomacy*, IB Tauris, London.

Jalsrot Vikas Sanstha 2012, *Ganges Strategic Basin Assessment: Consultative Workshop Report*, Report of workshop held in Lalitpur, Nepal, on 20 July 2012, published in Kathmandu, viewed 2 October 2016, www.jvs-nwp.org.np/sites/default/files/Consultative%20workshop%20report_GSBA.pdf

Jansky, L, Nakayama, M & Pachova, NI 2008, 'Introduction: From Domestic to International Water Security' in NI Pachova, M Nakayama & L Jansky (eds), *International Water Security: Domestic Threats and Opportunities*, United Nations University Press, United States of America, pp. 1–6.

Jeuland, M, Harshadeep, N, Escurra, J, Blackmore, D & Sadoff, C 2013, 'Implications of Climate Change for Water Resources Development in the Ganges Basin', *Water Policy*, vol. 15, pp. 26–50.

Jones, P 2008, 'South Asia: Is a Regional Security Community Possible?', *South Asian Survey*, vol. 15, no. 2, pp. 183–193.

Jones, P 2015, *Track Two Diplomacy in Theory and Practice*, Stanford University Press, Stanford.

Julien, F 2012, 'Hydropolitics Is What Societies Make of It (or Why We Need a Constructivist Approach to the Geopolitics of Water)', *International Journal of Sustainable Society*, vol. 4, no. 1/2, pp. 45–71.

Kayastha, RL 2001, 'Water Resources Development of Nepal: A Regional Perspective' in AK Biswas & JI Uitto (eds), *Sustainable Development of the Ganges–Brahmaputra–Meghna Basins*, Oxford University Press, New Delhi, pp. 122–44.

Keohane, RO & Nye, J 2001, *Power and Interdependence*, 3rd edn, Longman, New York.

Keskinen, M, Inkinen, A, Hakanen, U, Rautavaara, A & Niinioja, M 2014, 'Water Diplomacy: Bringing Diplomacy into Water Cooperation and Water into Diplomacy' in G Pangare (ed.), *Hydro Diplomacy: Sharing Water Across Borders*, Academic Foundation, New Delhi, pp. 35–40.

Kibaroglu, A, Brouma, AD & Erdem, M 2008, 'Transboundary Water Issues in the Euphrates–Tigris River Basin: Some Methodological Approaches and Opportunities for Cooperation' in NI Pachova, M Nakayama & L Jansky (eds), *International Water Security: Domestic Threats and Opportunities*, United Nations University Press, United States of America, pp. 223–251.

Kugelman, M 2012, 'Safeguarding South Asia's Water Security; The Twin Imperatives of Effective Transnational Water Arrangements and Internal Water Management' in L Powell & S Mittra (eds), *Perspectives on Water: Constructing Alternative Narratives*, Academic Foundation, New Delhi, pp. 177–195.

Kumar, M & Furlong, M 2012, 'Securing the Right to Water in India: Perspectives and Challenges', Our Right to Water, Blue Planet Project, viewed 2 October 2016, www.indiaenvironmentportal.org.in/category/34030/publisher/the-blue-planet-project/

Kurup, A 2010, 'Integrated Water Resource Management (IWRM): An Alternative Paradigm' in NS Mohan, S Routray & N Sashikumar (eds), *River Water Sharing: Transboundary Conflict and Cooperation in India*, Routledge, New Delhi, pp. 156–180.

Lahiri-Dutt, K 2008, 'Introduction' in K Lahiri-Dutt & RJ Wasson (eds), *Water First: Issues and Challenges for Nations and Communities in South Asia*, Sage, New Delhi, pp. xix–l.

Lahiri-Dutt, K & Wasson, RJ (eds) 2008, *Water First: Issues and Challenges for Nations and Communities in South Asia*, Sage, New Delhi.

Lang, DJ, Wiek, A, Bergmann, M, Stauffacher, M, Martens, P, Moll, P, Swilling, M & Thomas, CJ 2012, 'Transdisciplinary Research in Sustainability Science: Practice, Principles, and Challenges', *Sustainability Science: Bridging the Gap between Science and Society*, doi:10.1007/s11625-011-0149-x, vol. 7, no. 1, pp. 25–43.

Lang, W 1993, 'A Professional's View', in GO Faure & JZ Rubin (eds), *Culture and Negotiation: The Resolution of Water Disputes*, Sage, Newbury Park, CA, pp. 38–46.

Langford, M 2005, 'The United Nations Concept of Water as a Human Right: New Paradigm or Old Problems?', *International Journal of Water Resources Development*, vol. 21, no. 2, pp. 273–282.

London Water Research Group 2016, viewed 2 October 2016, www.lwrg.wordpress.com

Loorbach, D, Huffenreuter, RL, Frantzeskaki, N & Rotmans, J 2015, 'Managing Transitions to Sustainable Provision of Global Public Goods' in JW Evans & R Davies, (eds), *Too Global to Fail: The World Bank at the Intersection of National and Global Public Policy in 2025*, Directions in Development, doi:10.1596/978-1-4648-0307-9, World Bank, Washington DC, pp. 233–262.

Loucks, DP 2000, 'Sustainable Water Resources Management', *Water International*, vol. 25, no. 1, pp. 3–10.

Malhorta, P 2010, 'Water Issues between Nepal, India and Bangladesh: A Review of Literature', Special Report #95, Institute of Peace and Conflict Studies, New Delhi.

Matthew, R 2014, 'Foreword' in V Asthana & AC Shukla 2014, *Water Security in India: Hope, Despair, and the Challenges of Human Development*, Bloomsbury Academic, New York, pp. ix–xii.

Mayne, J 1999, Addressing Attribution through Contribution Analysis: Using Performance Measures Sensibly, Discussion Paper, June, Office of the Auditor General of Canada, Ottawa.
Mays, LW (ed.) 2007, *Water Resources Sustainability*, WEF Press, Alexandria, VI.
Mirumachi, N 2008, 'Domestic Issues in Developing International Waters in Lesotho: Ensuring Water Security amidst Political Instability' in NI Pachova, M Nakayama & L Jansky (eds), *International Water Security: Domestic Threats and Opportunities*, United Nations University Press, United States of America, pp. 35–60.
Mirumachi, N 2013, 'Securitising Shared Waters: An Analysis of the Hydropolitical Context of the Tanakpur Barrage Project between Nepal and India', *The Geographical Journal*, doi:10.1111/geoj.12029, vol. 179, no. 4, pp. 309–319.
Mirumachi, N 2015, *Transboundary Water Politics in the Developing World*, Routledge, Abingdon.
Mirumachi, N & Allan, JA 2007, 'Revisiting Transboundary Water Governance: Power, Conflict, Cooperation and the Political Economy', Proceedings from CAIWA International Conference on Adaptive and Integrated Water Management: Coping with Scarcity, Basel, Switzerland, 12–15 November 2007.
Mitra, D 2010, *Indo-Bhutan Relations: Political Process, Conflict and Crisis*, Academic Excellence, New Delhi.
Mitra, SK 2003, 'The Reluctant Hegemon: India's Self-perception and the South Asian Strategic Environment', *Contemporary South Asia*, vol. 12, no. 3, pp. 399–417.
Mittra, S 2014, personal communication, 3 June 2014, New Delhi.
Mohan, NS 2010, 'Locating Transboundary Water Sharing in India' in NS Mohan, S Routray & N Sashikumar (eds), *River Water Sharing: Transboundary Conflict and Cooperation in India*, Routledge, New Delhi, pp. 3–22.
Mohan, NS, Routray, S & Sashikumar, N (eds) 2010, *River Water Sharing: Transboundary Conflict and Cooperation in India*, Routledge, New Delhi.
Molden, D 2014, personal communication, 6 August 2014, Kathmandu.
Mollinga, P 2006, 'IWRM in South Asia: A Concept Looking for a Constituency' in K Athukorala, A Dixit & P Mollinga (eds), *Integrated Water Resources Management: Global Theory, Emerging Practice, and Local Needs*, Sage, New Delhi, pp. 21–37.
Mollinga, P 2008a, 'Foreword' in K Lahiri-Dutt & RJ Wasson (eds), *Water First: Issues and Challenges for Nations and Communities in South Asia*, Sage, New Delhi, pp. xiii–xvi.
Mollinga, P 2008b, 'Water, Politics and Development: Framing a Political Sociology of Water Resources Management', *Water Alternatives*, vol. 1, no. 1, pp. 7–23.
Mollinga, P & Gondhalekar, D 2014, 'Finding Structure in Diversity: A Stepwise Small-N/Medium-N Qualitative Comparative Analysis Approach for Water Resources Management Research', *Water Alternatives*, vol. 7, no. 1, pp. 178–198.
Mollinga, P, Meinzen-Dick, RS & Merrey, DJ 2007, 'Politics, Plurality and Problemsheds: A Strategic Approach for Reform of Agricultural Water Resources Management', *Development Policy Review*, vol. 23, no. 6, pp. 600–719.
Muller, M 2015, 'The "Nexus" as a Step Back Towards a More Coherent Water Resource Management Paradigm', *Water Alternatives*, vol. 8, no. 1, pp. 675–694.
Mukherjee, S 2001, 'Water Resource Development in Bhutan' in AK Biswas & JI Uitto (eds), *Sustainable Development of the Ganges–Brahmaputra–Meghna Basins*, Oxford University Press, New Delhi, pp. 145–162.
National Research Council of the National Academies 2012, *Himalayan Glaciers: Climate Change, Water Resources, and Water Security*, The National Academies Press, Washington DC.

Nexant SARI/Energy 2002, Regional Hydro-power Resources: Status of Development and Barriers (Bhutan), USAID-SARI/Energy Program, September, http://pdf.usaid.gov/pdf_docs/Pnada886.pdf

Nishat, A 2014, 'A Review of the Institutional Framework for Management of Water Resources in the Ganges–Brahmaputra–Meghna System' in G Pangare (ed.), *Hydro Diplomacy: Sharing Water Across Borders*, Academic Foundation, New Delhi, pp. 109–114.

Novel, S 2014, personal communication, 13 October 2014, Dhaka.

OECD 2005, Paris Declaration for Aid Effectiveness, OECD, Paris, France.

OECD 2008, Accra Agenda for Action, OECD, Accra, Ghana.

OECD 2012, Busan Partnership for Effective Development Cooperation, OECD, Busan, South Korea.

O'Lear, S, Briggs, CM & Denning, GM 2013, 'Environmental Security, Military Planning, and Civilian Research: The Case of Water', *Environment: Science and Policy for Sustainable Development*, vol. 55, no. 5, pp. 3–13.

Onta, IR 2001, 'Harnessing the Himalayan Waters of Nepal: A Case for Partnership on the Ganges Basin' in AK Biswas & JI Uitto (eds), *Sustainable Development of the Ganges–Brahmaputra–Meghna Basins*, Oxford University Press, New Delhi, pp. 100–121.

Pachova, N, Nakayama, M & Jansky, L 2008, 'National Sovereignty and Human Security: Changing Realities and Concepts in International Water Management' in NI Pachova, M Nakayama & L Jansky (eds), *International Water Security: Domestic Threats and Opportunities*, United Nations University Press, United States of America, pp. 289–295.

Pahl-Wostl, C 2015, *Water Governance in the Face of Global Change: From Understanding to Transformation*, Springer, London.

Pangare, G (ed.) 2014, *Hydro Diplomacy: Sharing Water Across Borders*, Academic Foundation, New Delhi.

Paranjpye, V 2010, 'Evolving a Negotiated Approach to Sharing of Transboundary Rivers' in NS Mohan, S Routray & N Sashikumar (eds), *River Water Sharing: Transboundary Conflict and Cooperation in India*, Routledge, New Delhi, pp. 121–139.

Patrick, MJ 2014, 'The Cycles and Spirals of Justice in Water-Allocation Decision Making', *Water International*, doi:10.1080/02508060.2013.863646, vol. 39, no. 1, pp. 63–80.

Patrick, MJ, Syme, GJ & Horwitz, P 2014, 'How Reframing a Water Management Issue Across Scales and Levels Impacts on Perceptions of Justice and Injustice', *Journal of Hydrology*, vol. 519, part C, pp. 2475–2482.

Patterson, JJ, Lukasiewicz, A, Wallis, PJ, Rubenstein, N, Coffey, B, Gachenga, E & Lynch, AJJ 2013 'Tapping Fresh Currents: Fostering Early-Career Researchers in Transdisciplinary Water Governance Research', *Water Alternatives*, vol. 6, no. 2, pp. 293–312.

Pearce, F 2006, *When the Rivers Run Dry: Water – The Defining Crisis of the Twenty-first Century*, Beacon Press, Boston.

Petersen-Perlman, JD & Wolf, AT 2015, 'Getting to the First Handshake: Enhancing Security by Initiating Cooperation in Transboundary River Basins', *Journal of the American Water Resources Association*, doi:10.1111/1752-1688.12348, vol. 51, no. 6, pp. 1688–1707.

Piper, K 2014, *The Price of Thirst: Global Water Inequality and the Coming Chaos*, University of Minnesota Press, Minneapolis.

Plummer, R 2005, 'A Review of Sustainable Development Implementation Through Local Action From an Ecosystem Management Perspective', *Journal of Rural and Tropical Public Health*, vol. 4, pp. 33–40.

Pohl, B, Carius, A, Conca, K, Dabelko, G, Kramer, A, Michel, D, Schmeier, S, Swain, A & Wolf, A 2014, *The Rise of Hydro-diplomacy: Strengthening Foreign Policy for Transboundary Waters*, adelphi research gemeinnützige GmbH, Berlin.

Powell, L & Mittra, S (eds) 2012, *Perspectives on Water: Constructing Alternative Narratives*, Academic Foundation, New Delhi.

Power, M 2009, 'The Commonwealth, "Development" and Post-colonial Responsibility', *Geoforum*, vol. 40, no. 1, pp. 14–24.

Prakash, A 2014, personal communication, 18 August 2014, Kathmandu.

Prasai, S 2014, personal communication, 7 July 2014, New Delhi.

Prasai, S & Surie, MD 2013, *Political Economy Analysis of the Teesta River Basin*, The Asia Foundation, New Delhi.

Prasai, S., and Surie, MD 2014, personal communication, 7 July 2014, New Delhi.

Prasai, S & Surie, MD 2015, 'Water and Climate Data in the Ganges Basin: Assessing Access to Information Regimes and Implications for Cooperation on Transboundary Rivers', *Water Alternatives*, vol. 8, no. 2, pp. 20–35.

Pun, SB 2013, 'World Bank's 2012 Ganges Strategic Basin Assessment: A View from Nepal', *Hydro Nepal*, vol. 12, pp. 6–12.

Qaddumi, H 2008, 'Practical Approaches to Transboundary Water Benefit Sharing', Working Paper 292, Overseas Development Institute, London.

Ramesh, S 2012, 'Revising the Draft National Water Policy', InfoChange, viewed 2 June 2013, www.infochangeindia.org/water-resources/analysis/revising-the-draft-national-water-policy.html

Ray, B 2008, 'Global Conventions and Regulations on International Rivers: Implications for South Asia' in K Lahiri-Dutt & RJ Wasson (eds), *Water First: Issues and Challenges for Nations and Communities in South Asia*, Sage, New Delhi, pp. 81–98.

Ray, P 2014, personal communication, 20 October 2014, Kolkata.

Riddell, RC 2014, Does Foreign Aid Really Work? An Updated Assessment, Devpolicy Discussion Paper no. 33, 1 March, Canberra, viewed 21 March 2015, http://papers.ssrn.com/sol3/papers.cfm?abstract_id=2409847

Rogers, P & Hall, AW 2003, Effective Water Governance, TEC Background Papers no. 7, Global Water Partnership, Stockholm.

Rollason, R 2012, 'Design Summary and Implementation Document: South Asia Water Initiative – Proposed Program and Approach', draft from 9 March 2012, AusAID, Canberra.

Routray, S 2010, 'The Water Sector in India: An Overview' in NS Mohan, S Routray & N Sashikumar (eds), *River Water Sharing: Transboundary Conflict and Cooperation in India*, Routledge, New Delhi, pp. 23–44.

Roy, D 2014, personal communication, 24 June 2014, New Delhi.

Royal Government of Bhutan & Bhutan Water Partnership 2003, *Bhutan Water Policy*, April 2003, Royal Government of Bhutan, Thimpu.

Sachs, JD 2015, *The Age of Sustainable Development*, Columbia University Press, New York.

Sadoff, C 2009, 'Ganges Strategic Basin Assessment (SBA)', Presentation given at The 4th Abu Dhabi Dialogue, 22–23 October 2009, Abu Dhabi, United Arab Emirates, viewed 19 March 2015, www.southasiawaterinitiative.org/sites/sawi/files/Sadoff%20Ganges.pdf

Sadoff, C & Grey, D 2002, 'Beyond the River: The Benefits of Cooperation on International Rivers' *Water Policy*, vol. 4, no. 5, pp. 389–403.

Sadoff, C, & Grey, D 2005, 'Cooperation on International Rivers', *Water International*, doi:10.1080/02508060508691886, vol. 30, no. 4, pp. 420–427.

Sadoff, C, Rao, H & team 2011, 'The Ganges Strategic Basin Assessment', presentation given at World Water Week 2011, 21 August, Stockholm, viewed 19 March 2015, www.worldwaterweek.org/documents/WWW_PDF/2011/Sunday/K23/Promoting-Cooperation-in-the-Ganges-Basin-through-Dialog/The-Ganges-Strategic-Basin-Assessment.pdf

Salame, L 2014, 'Working Towards Potential Conflict to Cooperation Potential' in G Pangare (ed.), *Hydro Diplomacy: Sharing Water Across Borders*, Academic Foundation, New Delhi, pp. 131–135.

Saravanan, VS 2008, 'Top-down or Bottom-up? Negotiating Water Management at the Local Level in South Asia' in K Lahiri-Dutt & RJ Wasson (eds), *Water First: Issues and Challenges for Nations and Communities in South Asia*, Sage, New Delhi, pp. 224–242.

Sayed, T 2012, Ganges Strategic Basin Assessment, Letter from World Bank, Nepal, to Mr. Surya Nath Upadhyay, Secretary General of Jalsrot Vikas Sanstha, Nepal, sent on 20 December 2012.

Seth, BL 2012, 'National Water Policy 2012 Silent on Priorities', 10 February, Down to Earth, viewed 2 October 2016, www.downtoearth.org.in/content/national-water-policy-2012-silent-priorities

Shah, RB 2001, 'Ganges–Brahmaputra: The Outlook for the Twenty-first Century' in AK Biswas & JI Uitto (eds), *Sustainable Development of the Ganges–Brahmaputra–Meghna Basins*, Oxford University Press, New Delhi, pp. 17–45.

Shakya, S 2009, *Unleashing Nepal*, Penguin, New Delhi.

Sharma, B 2014, 'Cleaning the Ganga, Step by Step', *The Hindu*, published online 12 July, viewed 23 July 2014, www.thehindu.com/opinion/op-ed/cleaning-the-ganga-step-by-step/article6105068.ece

Sharma, M 2012, *Green and Saffron; Hindu Nationalism and Indian Environmental Politics*, Permanent Black, New Delhi.

Siddiqui, S 2012, 'Water Use Efficiency in the Indian Context: A Regulatory and Institutional Mapping' in L Powell & S Mittra (eds), *Perspectives on Water: Constructing Alternative Narratives*, Academic Foundation, New Delhi, pp. 153–164.

Siddiqui, S 2014, personal communication, 20 June 2014, New Delhi.

Singh, R 2008, 'Trans-boundary Water Politics and Conflicts in South Asia: Towards Water for Peace', Centre for Democracy and Social Action, New Delhi.

Smith, MD 2010, 'Principles in Practice: Updating the Global Multi-Stakeholder Dialogue on Dams in 2010', *Water Alternatives*, vol. 3 no. 2, pp. 438–443.

Sneddon, C 2013, 'Water, Governance and Hegemony' in LM Harris, JA Goldin & C Sneddon (eds), *Contemporary Water Governance in the Global South: Scarcity, Marketization and Participation*, Routledge, Abingdon, pp. 13–24.

South Asia Water Initiative 2015a, 'Approaches and Objectives', viewed 11 January 2015, www.southasiawaterinitiative.org/SAWIApproach

South Asia Water Initiative 2015b, 'Strategic Alignment', viewed 11 January 2015, www.southasiawaterinitiative.org/SAWIAlignment

South Asia Water Initiative 2015c, 'The Abu Dhabi Dialogue (ADD)', viewed 16 March 2015, www.southasiawaterinitiative.org/node/11

South Asia Water Initiative 2015d, 'The Abu Dhabi Dialogue: 2007', viewed 28 April 2015, www.southasiawaterinitiative.org/node/97

South Asia Water Initiative 2015e, 'The Abu Dhabi Dialogue: 2006', viewed 26 April 2015, www.southasiawaterinitiative.org/node/89

South Asia Water Initiative 2015f, 'The Abu Dhabi Dialogue: 2008', viewed 28 April 2015, www.southasiawaterinitiative.org/node/91

South Asia Water Initiative 2015g, 'The Abu Dhabi Dialogue: 2009', viewed 28 April 2015, www.southasiawaterinitiative.org/node/96

South Asia Water Initiative 2015h, 'The Abu Dhabi Dialogue: 2010', viewed 28 April 2015, www.southasiawaterinitiative.org/node/93

South Asia Water Initiative 2015i, 'The Abu Dhabi Dialogue: 2012', viewed 28 April 2015, www.southasiawaterinitiative.org/node/95

South Asia Water Initiative 2015j, 'The Small Grants Program', viewed 28 April 2015, www.southasiawaterinitiative.org/node/12

South Asian Network on Dams, Rivers and People 2012, 'Not the Farmers, Not the Environment Draft National Policy 2012 Seems to Help Only Vested Interests', press release 2 February, South Asian Network on Dams, Rivers and People, New Delhi.

Stein, H 2008, *Beyond the World Bank Agenda: An Institutional Approach to Development*, The University of Chicago Press, Chicago.

Subedi, SP 2005, *Dynamics of Foreign Policy and Law: A Study in Indo-Nepal Relations*, Oxford University Press, New Delhi.

Surie, MD 2014, personal communication, 7 July 2014, New Delhi.

Swain, A 2002, 'The Nile Basin Initiative: Too Many Cooks, Too Little Broth', *SAIS Review*, vol. 22, no. 2, pp. 293–308.

Swain, A 2011, 'Challenges for Water Sharing in the Nile Basin: Changing Geo-politics and Changing Climate', *Hydrological Sciences Journal*, doi:10.1080/02626667.2011.577037, vol. 56, no. 4, 687–702.

Tamang, R 2009, 'Geopolitics and Shifts in Development Aid Policies: The Effects on Poverty in Nepal', *Journal of Sustainable Development*, vol. 2, no. 2, pp. 44–53.

Thakkar, H 2014, personal communication, 24 June 2014, New Delhi.

Thapa, AB 2009, 'Need to Review the West Seti Dam Design', *Hydro Nepal*, vol. 5, pp. 35–36.

Thomas, K 2012, 'Water Under the Bridge? International Resource Conflict and Post-Treaty Dynamics in South Asia', *South Asia Journal*, vol. 5, pp. 11–28.

Times of India 2012, 'India Rejects Aid from Britain, Says it is Peanuts', 5 February, viewed 25 May 2015, http://timesofindia.indiatimes.com/world/uk/India-rejects-aid-from-Britain-says-it-is-peanuts/articleshow/11769480.cms?referral=PM

Tiwary, R 2006, 'Conflicts over International Waters', *Economic and Political Weekly*, vol. 41, no. 17, pp. 1684–1692.

Treaty Between His Majesty's Government of Nepal and the Government of India Concerning the Integrated Development of the Mahakali Barrage Including Sarada Barrage, Tanakpur Barrage and Pancheshwar Project, signed 12 February 1996.

Treaty Between the Government of the People's Republic of Bangladesh and the Government of the Republic of India on Sharing of the Ganga/Ganges Waters at Farakka, signed 12 December 1996.

Uitto, JI, & Duda, AM 2002, 'Management of Transboundary Water Resources: Lessons from International Cooperation for Conflict Prevention', *The Geographical Journal*, vol. 168, no. 4, pp. 365–378.

UNESCO 2013, *Free Flow: Reaching Water Security Through Cooperation*, Tudor Rose, Paris.

United Nations 1997, *Convention on the Law of the Non-Navigational Uses of International Watercourses*, General Assembly Resolution A/RES/51/229, adopted by the General Assembly of the United Nations on 21 May 1997.

United Nations 2010, *The Human Right to Water and Sanitation*, General Assembly Resolution A/RES/64/292, adopted by the General Assembly of the United Nations on 28 July 2010.

United Nations 2012, *The Future We Want*. UN Resolution A/RES/66/288, adopted by the General Assembly on 27 July.

United Nations 2015a, *Zero Draft of the Outcome Document for the UN Summit to Adopt the Post-2015 Development Agenda*, viewed 25 July 2015, www.un.org/youthenvoy/2015/06/zero-draft-outcome-document-adopt-post-2015-development-agenda/

United Nations 2015b, *Transforming Our World: The 2030 Agenda for Sustainable Development*, General Assembly Resolution A/RES/70/1, adopted by the General Assembly of the United Nations on 25 September 2015.

United Nations Economic and Social Commission for Asia and the Pacific 2009, What is Good Governance? United Nations Economic and Social Commission for Asia and the Pacific, Bangkok.

United Nations University & United Nations Office of Sustainable Development 2013, Water for Sustainability: Framing Water within the Post-2015 Development Agenda, United Nations University Institute for Water, Environment and Health, UN Office of Sustainable Development and Stockholm Environment Institute.

United Nations Water 2013, International Year of Water Cooperation, viewed 2 October 2016, www.unwater.org/water-cooperation-2013/water-cooperation/en/

United Nations Water 2014, A Post-2015 Global Goal for Water: Synthesis of Key Findings and Recommendations from UN-Water, 27 January, viewed 2 October 2016, www.un.org/waterforlifedecade/pdf/27_01_2014_un-water_paper_on_a_post2015_global_goal_for_water.pdf

Upreti, BR 2004, 'Resource Conflicts and Conflict Resolution in Nepal', *Mountain Research and Development*, vol. 24, no. 1, pp. 60–66.

Uprety, K 2014, A South Asian Perspective on the UN Watercourses Convention, International Water Law Project Blog, published online 14 July, viewed 23 July 2014, www.internationalwaterlaw.org/blog/2014/07/14/dr-kishor-uprety-a-south-asian-perspective-on-the-un-watercourses-convention/

Vaidya, RA & Sharma, E (eds) 2014, *Research Insights on Climate and Water in the Hindu Kush Himalayas*, ICIMOD, Kathmandu.

Varady, RG, Gerlak, AK & McGovern, ED 2014, 'Hydrosolidarity and Its Place in International Water Diplomacy', in Pangare, G (ed.), *Hydro Diplomacy: Sharing Water Across Borders*, Academic Foundation, New Delhi, pp. 23–31.

Verghese, BG 2007, *Waters of Hope: Facing New Challenges in Himalaya–Ganga Cooperation*, India Research Press, New Delhi.

Warner, J 2004, 'Plugging the GAP: Working with Buzan: The Ilisu Dam as a Security Issue', Occasional Paper No 67, January 2004, SOAS Water Issues Study Group, School of Oriental and African Studies/King's College London.

Warner, JF & Meissner, R 2008, 'The Politics of Security in the Okavango River Basin: From Civil War to Saving Wetlands (1975–2002) – a Preliminary Security Impact Assessment' in NI Pachova, M Nakayama & L Jansky (eds), *International Water Security: Domestic Threats and Opportunities*, United Nations University Press, United States of America, pp. 252–286.

Warner, JF & Zawahri, N 2012, 'Hegemony and Asymmetry: Multiple-Chessboard Games on Transboundary River', *International Environmental Agreements: Politics, Law and Economics*, vol. 12, no. 3, pp. 215–229.

Werz, M & Conley, L 2012, 'Climate Change, Migration, and Conflict: Addressing Complex Crisis Scenarios in the 21st Century', Centre for American Progress, Heinrich Böll Stiftung: Washington DC.

Whitehead, J 2008, 'Submerged Voices and Transnational Environmentalism: The Movement against the Sardar Sarovar Dam' in K Lahiri-Dutt & RJ Wasson (eds), *Water First: Issues and Challenges for Nations and Communities in South Asia*, Sage, New Delhi, pp. 295–315.
Williams, MC 2003, 'Words, Images, Enemies: Securitization and International Politics', *International Studies Quarterly*, vol. 47, pp. 511–531.
Wirsing, R, Jasparro, C & Stoll, DC 2013, *International Conflict Over Water Resources in Himalayan Asia*, Palgrave Macmillan, New York.
Wohl, E 2012, 'The Ganga – Eternally Pure?', GWF Discussion Paper 1208, Global Water Forum, Canberra.
Wolf, AT 1998, 'Conflict and Cooperation Along International Waterways', *Water Policy*, vol. 1 no. 2, pp. 251–265.
Wolf, AT, Kramer, A, Carius, A & Dabelko, GD 2005, 'Managing Water Conflict and Cooperation', *State of the World 2005*, The Worldwatch Institute, Washington DC.
Wood, JR 2007, *The Politics of Water Resource Development in India: The Narmada Dams Controversy*, Sage, New Delhi.
World Bank, 2013, 'Water Resources Management: Sector Results Profile', 11 April, viewed 16 March 2015, www.worldbank.org/en/results/2013/04/15/water-resources-management-results-profile
World Bank 2014a, *Ganges Strategic Basin Assessment; A Discussion of Regional Opportunities and Risks*, Report no. 67668-SAS, The World Bank, Washington DC.
World Bank 2014b, *South Asia Water Initiative: Annual Report from the World Bank to Trust Fund Donors July 1, 2013–June 30, 2014*, The World Bank, Washington DC.
World Bank Group 2008, *Development and Climate Change: A Strategic Framework for the World Bank Group*, Report to the Development Committee, 12 October 2008, The World Bank Group, Washington DC.
World Commission on Dams 2000, *Dams and Development: A New Framework for Decision-Making*, Earthscan Publications, London.
World Commission on Environment and Development 1987, *Our Common Future*, Oxford University Press, New York.
Yoffe, S, Fiske, G, Giordano, M, Giordano, M, Larson, K, Stahl, K & Wolf, A 2004, 'Geography of International Water Conflict and Cooperation: Data Sets and Applications', *Water Resources Research*, vol. 40, no. 5, pp. 1–12.
Young, W 2014, personal communication, 20 May 2014, New Delhi.
Zartman, IW 1993, 'A Skeptic's View' in GO Faure & JZ Rubin (eds), *Culture and Negotiation: The Resolution of Water Disputes*, Sage, Newbury Park, CA, pp. 17–21.
Zeitoun, M & Allan, JA 2008, 'Applying Hegemony and Power Theory to Transboundary Water Analysis', *Water Policy*, vol. 10, supplement 2, pp. 3–12.
Zeitoun, M, Cascão, AE, Warner, J, Mirumachi, N, Matthews, N, Menga, F & Farnum, R 2016, 'Transboundary Water Interaction III: Contest and Compliance', *International Environmental Agreements: Politics, Law and Economics*, doi:10.1007/s10784-016-9325-x, pp. 1–24.
Zeitoun, M & Jägerskog, A 2009, 'Confronting Power: Strategies to Support Less Powerful States' in A Jägerskog & M Zeitoun (eds), *Getting Transboundary Water Right: Theory and Practice for Effective Cooperation*, Report no. 25, Stockholm International Water Institute, Stockholm.
Zeitoun, M & Mirumachi, N 2008, 'Transboundary Water Interaction I: Reconsidering Conflict and Cooperation', *International Environmental Agreements: Politics, Law and Economics*, vol. 8, no. 4, pp. 297–316.

Zeitoun, M, Mirumachi, N, & Warner, J 2011, 'Transboundary Water Interaction II: The Influence of "Soft" Power', *International Environmental Agreements: Politics, Law and Economics*, vol. 11, no. 2, pp. 159–178.
Zeitoun, M & Warner, J 2006, 'Hydro-hegemony – a Framework for Analysis of Transboundary Water Conflicts', *Water Policy*, vol. 8, pp. 435–460.
Zeitoun, M, Warner, J, Mirumachi, N, Allan, JA, Matthews, N, Cascao, A, McLaughlin, K 2013, 'Transboundary Water Justice: An Exploration of Social Justice and the Analysis of International Transboundary Water Interaction', Draft exploratory piece prepared in advance of Hydro-Hegemony 6: Transboundary Water Justice, London, 12–13 January 2013.
Zeitoun, M, Cascão, AE, Warner, J, Mirumachi, N, Farnum, R, Matthews, N 2014a, 'Transboundary Water Interaction III: Contesting Hegemonic Arrangements', Concept paper prepared for the Hydro-Hegemony 7 workshop, 10–11 May 2014, held in London by the University of East Anglia.
Zeitoun, M, Warner, J, Mirumachi, N, Matthews, N, McLaughlin, K, Woodhouse, M, Cascão, A & Allan, JA 2014b, 'Transboundary Water Justice: A Combined Reading of Literature on Critical Transboundary Water Interaction and "Justice", for Analysis and Diplomacy', *Water Policy*, vol. 16, pp. 174–193.
Zia, BK 2012, 'Bangladesh–India Relations, Challenges and Prospects', *Strategic Analysis*, doi:10.1080/09700161.2012.712368, vol. 36, no. 5, pp. 720–722.

Index

Abu Dhabi Dialogue (ADD) 5–6, 7, 12, 65, 82–4, 91, 95–6, 125–6, 141, 156, 164
agriculture 42, 44; *see also* irrigation
aid *see* international development aid
anti-dam movements 102; *see also* civil society
Asia Development Bank (ADB) 11
Asia Infrastructure Investment Bank (AIIB) 59
Australia 5, 6, 11, 59, 61, 66, 81, 83, 85, 86, 87–9, 92, 95, 123, 124

Bangladesh 4, 47, 54, 94, 98; demographics 3; domestic politics 37–8; transboundary water interactions with India 115–16, 162
benefit sharing policy 6, 18, 21, 26–8, 30–1, 47, 85, 97; in Ganges-Brahmaputra problemshed 50–1
Bhutan 4, 121; agriculture 55; domestic politics 39; hydropower potential 52; transboundary water interactions with India 115–16, 162
bilateralism *see* water governance; bilateral
Brahmaputra River; Great Bend of 14
BRICS Bank *see* New Development Bank

Central Asia 34
China 12–14, 39, 62, 65, 93, 94, 149
civil society 101–3, 153; *see also* anti-dam movements
climate change 41–3, 60, 74, 158–9; adaptation 5, 87–8, 96
collective action problems 73, 162; *see also* non-traditional security threats
command and control paradigm 4, 33–4, 35–6, 55, 58, 67, 74, 78–80, 153–5, 160
complex adaptive systems 2, 28–31; water as 14–15, 71–2, 100, 130, 137–8, 150, 165–6, 167–8; *see also* water governance

dams *see* multi-purpose dams
deliberative governance 6, 16–17, 23, 70, 85, 93, 96, 147–8; *see also* good governance; Track II dialogue
discursive governance *see* deliberative governance

energy security 67; *see also* non-traditional security threats
environmental flows 80
environmental security 72, 73; *see also* non-traditional security threats

Farakka Barrage 47, 51–2, 111, 130, 151
flooding 55–7; Kosi River 2008 55; management of 33, 97, 159
food security 44, 67, 75; *see also* non-traditional security threats
foreign aid *see* international development aid

Ganges Strategic Basin Assessment (GSBA) 27, 43, 45–6, 48–50, 53, 55–6, 57, 75, 71, 84, 91, 91–2, 97–100, 112, 132, 155, 160
Ganges Waters Treaty 47, 120, 130
Ganges-Brahmaputra problemshed 2–4, 12–14, 32–44, 157–9, 159–60; colonial rule 33; demographics 3, 40–1, 158; domestic politics 36–9, 126, 144; history 4, 144
Gesellschaft für Internationale Zusammenarbeit (GIZ) 11
glacial melt 42, 43, 87
Global Water Partnership (GWP) 11

good governance 65, 68–72, 85–6; of water 67, 147–8; *see also* deliberative governance
Green Revolution 32–4, 53
groundwater 34, 46–7, 50

hegemonic concepts 67–8
human security 3; *see also* non-traditional security threats
hydro-diplomacy 5, 19, 29–31, 40, 73–4, 81, 85–6, 100, 101, 113–14, 132, 148–9
hydro-hegemony 23, 26, 109, 116–17, 131; countering 120–1, 145–6; *see also* India; as hydro-hegemon
hydro-sovereignty 30; *see also* sovereignty
hydropower 3, 50–3, 154–5

India 4, 66, 94, 95–6, 98; as hydro-hegemon 66, 96, 105, 109, 111–12, 115–21, 117–18, 127–8, 144–6, 152–61; demographics 3; domestic politics 37; inter-state water conflicts 36–7, 110; National River Linking Project 36, 54; National Water Policy 2012 48, 53, 96, 112, 119; Treaties of Friendship 119–20; water management 54
Indus basin 60, 129
Indus Waters Treaty (IWT) 64, 65, 129
Integrated Water Resources Management (IWRM) 6, 68, 87, 92, 124, 125
Interlinking Rivers Project *see* India; National River Linking Project
international actors 10–12; role in negotiation 19–21
International Centre for Integrated Mountain Development (ICIMOD) 11, 83, 89, 90, 94, 112, 124, 153
international development aid 7, 136–42, 167; transformational change 138–41
International Union for the Conservation of Nature (IUCN) 11, 92
International Water Management Institute (IWMI) 11
irrigation 33–4, 53–4, 159; *see also* agriculture

Japan 11, 61

knowledge sharing 89–90, 167; lack of in South Asia 110–14; limitations of 123–5
Kosi River flood 2008 *see* flooding; Kosi River 2008

Mahakali Treaty 15, 109, 111
Mekong basin 12, 17
Millennium Development Goals (MDGs)64, 75–6; *see also* Sustainable Development Goals (SDGs)
Monsoon 3, 42, 46, 87
multi-purpose dams 4, 46–50, 62, 159, 160; *see also* hydropower
multilateralism *see* water governance; multilateral

Narmada dams controversy 64; *see also* anti-dam movements
National River Linking Project *see* India; National River Linking Project
nationalism 102, 104–6, 109; *see also* sovereignty
negotiation in international disputes 19–21; effect of culture on 21–3; effect of power asymmetry between parties on 22–3
Nepal 4, 98; dams 47; demographics 3; domestic politics 38–9; hydropower potential 3, 47, 50–2, 97, 159; transboundary water interactions with India 47, 87, 109, 115–16
New Development Bank 66
NGOs 103
Nile basin 12
Nile Basin Initiative (NBI) 5, 7, 83, 126, 149, 156
non-traditional security threats4, 25, 40, 67, 72–4, 86–7, 107; in South Asia 59–60, 161; *see also* collective action problems
non-zero-sum outcomes *see* zero-sum outcomes
Norway 6, 11, 61, 81, 83, 88, 95

Pakistan 12; *see also* Indus Waters Treaty (IWT)
positive-sum outcomes *see* zero-sum outcomes
problemshed 2, 12–14, 125–6, 149–51, 167; Ganges-Brahmaputra 2, 12–14, 40–1

river basin organisations (RBOs) 152–3

Sardar Sarovar *see* Narmada dams controversy
securitisation of water 12–14, 107–10
sedimentation 159–60

South Asia Association for Regional Cooperation (SAARC) 105, 121
South Asia Water Initiative (SAWI) 5–6, 11, 17, 27, 81, 82–4, 85, 89, 91–100, 124, 125, 127, 139, 140, 145, 149, 164
sovereignty 91, 104–6; *see also* hydro-sovereignty; nationalism
supply side water management 41, 43–4, 48, 55, 153–5; *see also* command and control paradigm
sustainable development 77–8, 88–9
Sustainable Development Goals (SDGs) 69, 76, 77–8; *see also* Millennium Development Goals (MDGs)

Tanakpur Barrage 109
Teesta River 38, 111, 151
Tehri dam 62
Track II dialogue 16–17, 18, 19, 23–5, 29, 86, 92, 125, 127, 163, 164, 165, 167; limits of 133–5; *see also* deliberative governance
transboundary water interactions 2, 4, 15–16, 18–19, 25–6, 30, 67, 101–14, 160–2; negative 101–14; positive15, 18, 26, 27–8, 45–8, 78, 129, 146, 151, 162–5

UN Watercourses Convention 146
United Kingdom (UK) 5, 6, 11, 59, 61, 66, 81, 83, 85, 87, 88, 89, 92, 95

water conflicts 5, 16, 19, 31, 35, 81, 86, 107, 110, 128–31, 131–2, 165; caused by culture of secrecy 111, 114; in Ganges-Brahmaputra problemshed 5, 6, 14, 60, 107; non-binary nature of 73; resolution of 148; sub-state 130
water cooperation *see* transboundary water interactions; positive
Water Diplomacy Framework 29
water governance 1–2, 14, 67, 68–72, 85–6, 129; bilateral 4, 118–20, 160–1; multilateral 104–6, 145; *see also* command and control paradigm; complex adaptive systems; water as; good governance; of water
water justice 129, 131, 132–3, 147–8
water scarcity 35–6, 71, 72; narrative of 153–5
water security 67, 68, 71, 75–6, 108, 164; *see also* non-traditional security threats
water sharing treaties 15
water storage *see* multi-purpose dams
West Seti dam 52
World Bank 5, 7, 11, 35, 36–7, 53, 59, 60–2, 63, 64, 74, 83, 84, 86, 87, 91–100, 99, 124, 136, 139–40, 152–3, 155, 160
World Commission on Dams (WCD) 42, 62–4
World Wildlife Fund (WWF) 11

zero-sum outcomes 36, 106–7

For Product Safety Concerns and Information please contact our EU
representative GPSR@taylorandfrancis.com
Taylor & Francis Verlag GmbH, Kaufingerstraße 24, 80331 München, Germany

www.ingramcontent.com/pod-product-compliance
Ingram Content Group UK Ltd.
Pitfield, Milton Keynes, MK11 3LW, UK
UKHW022126190926
475702UK00013B/1680

* 9 7 8 0 3 6 7 3 5 2 2 3 3 *